MONOGRAPHS ON
STATISTICS AND APPLIED PROBABILITY

General Editors

D. R. Cox and D. V. Hinkley

Probability, Statistics and Time
M. S. Bartlett

The Statistical Analysis of Spatial Pattern
M. S. Bartlett

Stochastic Population Models in Ecology and Epidemiology
M. S. Bartlett

Risk Theory
R. E. Beard, T. Pentikäinen and E. Pesonen

Residuals and Influence in Regression
R. D. Cook and S. Weisberg

Point Processes
D. R. Cox and V. Isham

Analysis of Binary Data
D. R. Cox

The Statistical Analysis of Series of Events
D. R. Cox and P. A. W. Lewis

Queues
D. R. Cox and W. L. Smith

Stochastic Abundance Models
E. Engen

The Analysis of Contingency Tables
B. S. Everitt

Finite Mixture Distributions
B. S. Everitt and D. J. Hand

Population Genetics
W. J. Ewens

(Full details concerning this series are available from the Publishers)

Applications of Queueing Theory

SECOND EDITION

G. F. NEWELL

Professor of Transportation Engineering
University of California, Berkeley

LONDON   NEW YORK

CHAPMAN AND HALL

First published 1971 by
Chapman and Hall Ltd
11 New Fetter Lane, London EC4P 4EE
Second edition published 1982
Published in the USA by
Chapman and Hall
733 Third Avenue, New York NY 10017

ISBN 0 412 24500 0

British Library Cataloguing in Publication Data

Newell, G. F.
    Applications of queueing theory.—2nd ed.—
    (Monographs on statistics and applied probability)
    1. Queueing theory
    I. Title   II. Series
    519.8′2   TS7.9

    ISBN 0–412–24500–0

Library of Congress Cataloging in Publication Data

Newell, G. F. (Gordon Frank), 1925–
    Applications of queueing theory.

    (Monographs on applied probability and
statistics)
    1. Queueing theory. I. Title. II. Series.
    T57.9.N48   1982   519.8′2   82-4423
    ISBN 0–412–24500–0   AACR2

Contents

Preface to the first edition

The literature on queueing theory is already very large. It contains more than a dozen books and about a thousand papers devoted exclusively to the subject; plus many other books on probability theory or operations research in which queueing theory is discussed. Despite this tremendous activity, queueing theory, as a tool for analysis of practical problems, remains in a primitive state; perhaps mostly because the theory has been motivated only superficially by its potential applications. People have devoted great efforts to solving the 'wrong problems.'

Queueing theory originated as a very practical subject. Much of the early work was motivated by problems concerning telephone traffic. Erlang, in particular, made many important contributions to the subject in the early part of this century. Telephone traffic remained one of the principle applications until about 1950. After World War II, activity in the fields of operations research and probability theory grew rapidly. Queueing theory became very popular, particularly in the late 1950s, but its popularity did not center so much around its applications as around its mathematical aspects. With the refinement of some clever mathematical tricks, it became clear that exact solutions could be found for a large number of mathematical problems associated with models of queueing phenomena. The literature grew from 'solutions looking for a problem' rather than from 'problems looking for a solution.'

Mathematicians working for their mutual entertainment will discard a problem either if they cannot solve it, or if being soluble it is yet trivial. An engineer concerned with the design of a facility cannot discard the problem. If it is trivial, he should recognize it as such and do it. If he cannot solve it correctly, then he must do the best he can. The practical world of queues abounds with problems that cannot be solved elegantly but which must be analysed nevertheless. The literature on queues abounds with 'exact solutions,' 'exact bounds,' simulation models, etc.; with almost everything except common sense methods of 'engineering judgment.' It is no wonder that engineers resort to using formulas which they know they are using incorrectly, or run to the computer even if they need only to know something to within a factor or two.

In the last 15 years or so, I have suffered many times the frustration of failing to solve elegantly what appeared to be a straightforward practical queueing problem, subsequently to discover that I could find very accurate approximations with a reasonable effort, and finally that I could obtain some crude estimates with almost no effort at all. There is no reason why students should suffer the same way. They should benefit from the mistakes of others and learn to do things in a sensible way, namely in the opposite order.

The following is an attempt to turn queueing theory around and point it toward the real world. It is, in essence, the fourth evolution of a series of lecture notes written for a course entitled 'Applications of Queueing Theory to Transportation.' The relevance of the subject to transportation, rather than to other possible fields of application, derives mostly from the fact that the course was given primarily for transportation engineering students and in a department of transportation engineering. The students had a diverse background, but the majority were graduate students with an undergraduate training in civil engineering. Most had just completed a one-quarter introductory course in probability theory at the level of Paul L. Meyer, *Introductory Probability and Statistical Applications* (Addison-Wesley, 1965) and were taking, concurrently, an introductory course in mathematical statistics. Most would not have had a course in advanced calculus and many would have forgotten much of their elementary calculus (students with a strong formal mathematics background usually had just as much difficulty with some of the graphical techniques as the engineering-oriented students had with the mathematics).

Whereas most of the queueing literature deals with equilibrium distributions of queue length, the main emphasis here is on time-dependent behavior, particularly rush hours in which the arrival rate of customers temporarily exceeds the service rate. The reason for this is that these are the situations which usually create large queues, and the most important practical problems are those in which the size of the queue is really a cause for concern. It turns out that these are among the simplest problems to solve approximately, but are so difficult to solve exactly that no one has yet solved a single special case, at least not in a form suitable for computation.

The techniques emphasized here are mainly 'fluid approximations' and 'diffusion approximations.' The former employs mostly graphical methods; the latter involves some elementary properties of partial differential equations but otherwise uses only a mixture of graphical methods and elementary analysis. Nowhere is use made of generating functions, characteristic functions, or Laplace transforms, which are the standard tools of analysis in conventional queueing theory methods.

No attempt is made here to construct any bibliography except for an occasional reference in the text to some particular paper. Although most of the methods described here have appeared in the literature before in the analysis of special problems, there does not appear to have been any systematic treatment of approximate methods in queueing theory. Some things here may be 'original' in the sense that no one has used a particular mathematical trick to solve a particular problem, but the techniques used are all basically very old, having been used in physics or engineering to solve other types of problems, long before anyone heard of 'queueing theory.'

Except for a short chapter on 'Equilibrium distributions' (Chapter 5), there is very little overlap between what is given here, and what is presented in other books on queueing theory. Although what follows is self-contained, it is not intended as a substitute for the more conventional treatments, but rather as a supplement to them. For further study of the more conventional aspects, it is recommended that a student read D. R. Cox and W. L. Smith, *Queues* (Chapman and Hall, 1961), for a very concise introduction; A. M. Lee, *Applied Queueing Theory* (Macmillan, 1966), for some interesting case histories of attempts to apply queueing theory to practical problems; and J. Riordan, *Stochastic Service Systems* (John Wiley, 1962) or N. V. Prabhu, *Queues and Inventories* (John Wiley, 1965), for a more complete introduction to the typical literature.

It is a pleasure to acknowledge the cooperation and assistance of the transportation engineering students at Berkeley who struggled with me through three preliminary versions of the class notes upon which this book is based. One of these, Brian Allen, was kind enough to help correct the final proofs. The typing and retyping of the notes was done by Phyllis De Fabio.

I would also like to express my thanks to Dr Arnold Nordseick and Professor Elliott Montroll who helped me, as a graduate student and post-doctoral fellow many years ago, to develop some of the attitudes which have influenced this book.

Special thanks go to my wife Barbara, who must endure the lonely life of a scientist's wife, and to my parents, who patiently guided me through my youth.

Berkeley, June 1970   G. F. Newell

Preface to the second edition

More than ten years have passed since the first edition of this book was published. It is interesting to look back now to see what has evolved during that time, and to look closely at the major improvements of this second edition.

The first edition had a title that promised 'applications' but the book contained only a few examples plus some hints on how one might attack certain applied problems, and the 'queueing theory' was not that which several generations of applied probabilists had developed. The style of the first edition, however, needs no defense; the 'theory' has, in fact, been applied to the analysis of a wide variety of practical problems which could not be solved by traditional methods of queueing theory and will continue to be used as a practical tool.

The first edition was essentially the lecture notes of a course which had evolved over a span of about four years. Variations of this course have been given almost every year since, but as it evolved further, more and more applications of the deterministic approximations were added until they consumed more than half of the course. For lack of time, the diffusion approximations were gradually squeezed out.

Although the mathematics of the deterministic approximations is elementary in the sense that it involves only graphs, algebra, and calculus, it requires skill and ingenuity to apply. Students certainly do not consider it easy. Without question, however, these deterministic approximations have found application to a much wider range of practical problems than the stochastic theory simply because the stochastic analysis of even the simplest systems which involve several servers or customer types is too tedious to be of much practical value.

Since a textbook must, of necessity, treat only simple illustrations which can be described in one or two lectures, this second edition still gives only some hints as to how one can analyze more complex problems. It is certainly inappropriate to try to describe in detail how, for example, these methods can (and are) used for the analysis of traffic signal synchronization, production line design, bus dispatching policies, etc. The difficulty in these more complex applications comes not so much in the derivation of formulas for delays, queue lengths, etc. as in the interpretation of the results which typically contain many parameters. Each area of application involves special understanding of what questions one is trying to answer.

Although the text itself contains few references, I have added a bibliography of some of the applications of deterministic queueing with which I am familiar (mostly in the area of transportation engineering).

Even though most of the diffusion theory was eliminated from the original course, much of my own research during the last ten years has been in the area of stochastic approximations. When I started to write the present revision, I envisaged presenting an approximate stochastic version of most of the models described under the deterministic theory. I even taught an 'advanced' course as a means of testing some of the revised notes on stochastic approximations. This course, however, started from the beginning, and since most students of queueing theory are unfamiliar with properties of partial differential equations, this course barely covered some of the basic qualitative features of single-server queues. Much of what I had hoped to do must wait for some future occasion.

The first edition has been almost entirely rewritten, but the chapter titles remain almost the same, as does the general philosophy and style. The rejection of traditional approaches to queueing theory is perhaps even more emphatic.

Chapter 1 has changed little except that some notation has been revised and some other types of graphical representatives are discussed. The first problem set now introduces a hand simulation of a 'random walk' queue which, with little effort, gives students some preliminary feeling for the magnitude of statistical fluctuations. These simulations are very helpful in illustrating some of the effects discussed in later chapters.

The introduction to Chapter 2 on fluid approximations now discusses more thoroughly the qualitative features of some real queueing situations and questions of concern to engineers designing service systems. Two typical types of systems for which deterministic approximations are particularly useful are then analyzed, the rush hour with a steady service rate and a system with interrupted (pulsed) service (traffic signals and buses).

Chapter 3, describing the behavior of systems with several servers and/or customer types, has been considerably expanded to emphasize the benefits derived from drawing graphs of the cumulative arrivals and departures of anything which satisfies a conservation principle. Some of the illustrations are conventional (tandem queues or multiple-channel server queues) but many are designed simply to illustrate the art of modelling simple systems. The problem set contains other illustrations many of which are derived from real applications. It is the material in this chapter that has displaced much of the stochastic approximations in the course I have taught because this is the type of analysis that has proven to be most useful in the design of real systems. This chapter has become the main focus of the course in recent years.

The introductory chapter on stochastic models has been extensively rewritten to emphasize the theme that stochastic models should be chosen to represent the actual behavior of real systems not just to yield mathematically convenient formulas. It describes typical qualitative properties of arrival and departure processes, culminating in the argument that if one cannot find a convenient 'exact' model, one can usually evaluate the things one really wants simply by constructing a few hypothetical realizations of the cumulative arrivals and departures.

Chapter 5 on equilibrium distributions now includes a simple dimensional argument giving the typical magnitude of equilibrium queue lengths and relaxation times and discusses the question of how rapidly the traffic intensity can change if the queue distribution is to stay close to the equilibrium distribution. In the first edition these issues were postponed until Chapter 6 and obtained as a result of a rescaling of the diffusion equation (which makes the argument unnecessarily obscure). This chapter also contains a few more examples of traditional equilibrium queueing problems than the first edition, although the reader is still referred to other texts for a more detailed introduction to conventional methods.

Chapter 6, which is entirely new, deals with systems in which customers seldom interact because the server has a sufficiently large number of channels and/or the arrival rate is low. It includes the standard infinite-channel systems and loss systems but also some approximations for queueing systems in which customers are delayed only rarely. The behavior of queueing systems under light traffic has, for some reason, received little attention in the queueing theory literature. If, however, one can describe the system behavior for both light and heavy traffic, it requires little imagination to guess how the system would behave for intermediate traffic (where exact results may be difficult to obtain). Perhaps this chapter will inspire further research on low traffic approximations.

Chapters 7, 8, and 9 are devoted to diffusion approximations. Chapter 7 starts with a fairly general stochastic process in two (or more) dimensions having the property that the state of the system changes by only a small amount in a short time. This is then specialized to treat properties of the joint arrival and departure processes, diffusion equations with state-independent coefficients, boundary conditions, and one-dimensional equations for the queue distribution. Chapter 8 deals with equilibrium and transient queue behavior for constant arrival and service rates, while Chapter 9 treats time-dependent queues, particularly the stochastic version of the rush hour and pulsed service problems introduced in Chapter 2.

Considerable work has been done in recent years on various queueing systems; multiple-channel servers, tandem queues and some more general networks of service systems. Although I had, at one time, planned to add perhaps two more chapters dealing with some of these results, I have (temporarily) abandoned the attempt because it would take too much time to put much of this in proper perspective. The complexity of the results in the existing literature is way out of proportion to its usefulness (including my own research). It will be some time before I can sift out from this material that which might be appropriate for an introductory book, but maybe there will be a third edition some day. Certainly the most difficult task in the analysis of any real system is the collection of relevant data to describe what is happening and the existing literature on queueing theory gives very little assistance to an engineer in deciding what to measure and how to interpret the results.

It is a pleasure to acknowledge the help I have received from the many students who have noted errors in preliminary revisions of these notes, suffered through unclear expositions, and tried to solve ill-posed problems. I have learned the most, however, from those students and colleagues who have actually collected data and analyzed real problems, particularly Van Olin Hurdle, now at the University of Toronto, who is a master at the use of graphical techniques.

Phyllis DeFabio has continued to type most of the multiple versions of notes from which this book derives.

CHAPTER 1

Introduction

1.1 Nature of the subject

Queueing theory is concerned, generally, with the mathematical techniques for analyzing the flow of objects through some network. The network contains one or more locations at which there is some restriction on the times or frequencies at which the objects can pass. A conservation principle applies; the objects do not disappear or disintegrate. Any object which cannot immediately pass some restriction is stored in some real or fictitious reservoir until it can. As long as there are objects in the reservoir waiting to pass, the facility will pass them as rapidly as the restriction will permit.

The objects could be anything which move from place to place (and satisfy a conservation principle), people, cars, water, money, jobs to be done, etc. The restrictions could be a service facility for people, a highway bottleneck for cars, a valve regulating the flow of water, a rule for money transactions, a finite labor supply for work to be done, or a finite speed with which a computer can handle calculations to be done.

One could consider all sorts of networks, but we will be concerned here mostly with the rather simple geometry in which all objects flow along some channel and all pass through the same restrictions as illustrated schematically in Fig. 1.1(a). In many (perhaps most) real systems, however, the objects are not identical. Although these objects may differ in many ways (color, size, name, etc.) the typical characteristics of these objects which are relevant to queueing analysis are:

(a) Different objects may take different lengths of time to pass the restrictions (there are long jobs and short jobs).
(b) Delays to different objects may be worth different amounts of money (to delay an aircraft carrying 400 passengers costs more than to delay a private aircraft).

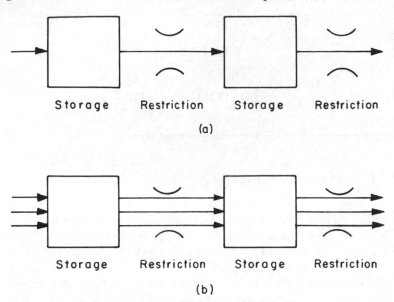

Figure 1.1 *Schematic picture of the flow of objects along a channel*

If one decomposes the objects into several categories, it is usually assumed that a conservation principle applies separately to each object type. Objects cannot disappear; nor can they change identity (a long job remains a long job, a commercial aircraft remains a commercial aircraft). A more realistic schematic picture of the system would be as shown in Fig. 1.1(b) with many streams passing the same restrictions. The restriction is usually a collective one restricting the rate at which objects can pass in various proportions.

   The ultimate practical purpose of any theory is to make predictions of what will happen in some experiment that one has not yet done. The purpose of queueing theory is to provide a mechanism for predicting how some hypothetical or proposed system will behave. Sometimes one wishes to design a completely new facility where there was none before and would like to compare the predicted performances of various proposed systems. In this case one must usually make conjectures about the arrival rates of various objects and the consequences of various restrictions (facilities). The more common problem, however, is one in which a facility already exists. One can make observations on its present behavior; but, from these observations, one would like to predict how the system would behave if certain changes occur. The change might be an increased demand (arrival rate) as projected for some future time or it might be an improvement in the service rate of some facility or a change in strategy for sequencing the service of different object types.

The same type of mathematical techniques apply to a very wide variety of flow systems, but the measures of performance or goals may be quite different for different systems. In queueing theory one typically associates an implied cost with any delay and also a cost for providing a higher service rate at any restriction. The usual problem is to compare delays (and operating costs) for systems with different service components or strategies. Some typical systems of this type are:

(a) Objects move along a production line on which various tasks are performed at various rates but there is a penalty for storage of unfinished products. One might be able to decrease this storage cost by shifting some labor.
(b) Cars move along a highway having certain bottlenecks (such as traffic intersections) and there is an inconvenience associated with waiting. One might be able to decrease the delays by appropriate adjustment of the signal timing.
(c) Patients wish to enter a hospital which can handle only finitely many patients at a time, but delays may cause serious consequences, more so to some patients than to others.

The mathematical models of 'inventory theory' are quite similar; objects pass from a supplier to a reservoir to a customer. The objectives and strategies, however, are quite different from the above. There is usually a high penalty for an empty inventory (or a queue of unfilled orders). The strategy is to regulate the input (reorder stock) rather than the flow out of the inventory. In the 'theory of dams' one has an input (rain) to a reservoir and an output (consumption) but the regulation or strategy is applied to the output rather than the input. In insurance, gambling, banking, etc., one is concerned with the flow of money. The money is, in fact, only some numbers on an account book, but the rules of transfer are the same as if it were something physical. There is a flow of money into an account (an investment rate), and a flow out; and a resulting storage (balance). The strategy now may involve regulation of either the input or the output or both with potentially a rather complex set of objectives. In the university one has a flow of students and faculty into and out of the system with a resulting population of each. There is a conservation principle: what comes in must go out, one way or another.

Since there is such a wide variety of possible applications of queueing theory or related theories, it is rather difficult to agree on some common terminology. Much of the conventional terminology has evolved from the following hypothetical system. The objects which move from place to place are called customers, which one typically imagines to be people. They arrive at some service point (a bank counter, a taxi stand, a highway intersection) at certain specified times. The service facility (the restriction) requires some time to serve each customer but is capable of serving only finitely many at a time (possibly just one). If customers arrive faster than the facility can serve them, they must wait in a queue (the reservoir).

Typically, both the customer arrival times and the service times are assumed to follow some specified stochastic behavior. One wishes to relate the delays to the customers, and the number of customers in the queue to the given properties of the arrival and service. In practical applications one usually wishes further to compare the operation of several possible modes of operation with respect to its type of service, cost, etc. Should there, for example, be a single queue for all bank tellers or separate queues for each?

In most of the following descriptions, we will also use the terms customer, server, and queue (except when it is clearly inappropriate) even though the terms object, restriction, and reservoir are more suggestive of the wide range of physical systems to which the same mathematics will apply.

What complicates the mathematical modelling of most real systems is that repetition of an experiment 'under identical conditions' does not usually yield exactly the same results every time. To make predictions of future behavior it is, generally, necessary to postulate some stochastic model and to estimate probabilities for certain events, i.e., fractions of times in which various events would happen over many repetitions of the observations. Unfortunately, in most applications of queueing theory, the observed properties of the stream of objects passing any point do not conform to any mathematically 'simple' stochastic model. To describe how the system behaves under repetitions of an experiment one must actually repeat the experiment to see what happens. It is usually rather dangerous to speculate on what would happen if the system were consistent with some hypothetical model. From repeated observations on an existing system one must, however, still make conjectures as to how the system would behave if certain changes were made in the system.

1.2 Mathematical and graphical representations of events

Since a study of any queueing system should start from some (real or implied) experimental observations, let us imagine that we station observers at various points in the system. For each service point we might place one observer just upstream of the server to record the times and identity of each customer that passes him. If customers travel with a finite speed and the queue has a positive physical length, we might ask this observer also to convert his observations into the times at which the customers would have reached the server if there were no queue (or if the queue occupied no space). We place a second observer at the server to record the times and identity of customers entering the server and possibly a third observer just downstream of the server to record the times at which customers leave the server (and their identity).

We will assume that at time 0, when the observations begin, the system is empty. If it is not, we can imagine that the system was empty for time $t < 0$ (whether it was or not) but that each observer records arbitrarily that any customer already downstream from him at $t = 0$ passed him at $t = 0$.

One could also imagine that the first observer assigns labels to each customer that passes him (for example, a number) and he asks the customer to keep the label with him at all times. All customers are now different by virtue of their labels (if not for other reasons) and each customer individually satisfies a conservation principle in that he does not disappear or change identity during the period of observations.

Since one may be interested in the possibility and possible consequences of the fact that customers might interchange positions in the queue or server (they pass each other), such a labeling will permit the downstream observers to detect any such rearrangement. We would, however, like to make a distinction between customers which differ only by virtue of having different labels and those which differ in some more significant way (the delay time of one is worth more than another or one is known to require a longer time in service) which may be relevant in selecting some service priorities. If it is relevant to treat the customer arrival as the superposition of several identifiably different streams, each satisfying a conservation principle, then we will ask each observer to record separately the times at which customers of each category pass.

For now, we will consider only the observations associated with customers within the same category, as if those of other categories (if any) were not there, or not observed.

If the first observer numbers the customers (of the same category) consecutively and assigns the numbers as labels let

$$0 \le t_1 \le t_2 \le \ldots$$

represent the time at which customers 1, 2, . . . arrive at the server or would arrive at the server if the queue occupied no space. These times can be represented graphically as a sequence of points on the real line as in Fig. 1.2(a). It is more convenient, however, to represent these data by a graph of a function $A(t)$ which, for each t, represents the cumulative number of arrivals to time t:

$$A(t) = \text{number of } t_j \text{ with } t_j \le t. \qquad (1.1)$$

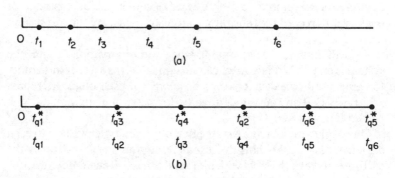

Figure 1.2   *Representation of arrival and departure times by points on a line*

This is a step function which increases by one at each time t_j as shown in Fig. 1.3.

One immediate advantage of this representation is that we will also have occasion to analyze arrivals and departures of quantities other than *numbers* of customers; for example, the cumulative value of products or the cumulative amount of work to be done. If the quantity in question is also conserved (what comes in must go out), then it is easy to generalize (1.1) to

$A(t) = $ cumulative quantity (or number) to arrive by time t.

(1.1a)

This is also a monotone nondecreasing function of t, but it is not necessarily integer valued. It may or may not be a step function depending upon whether or not the arrivals are discrete. If they are discrete, the steps need not be equal.

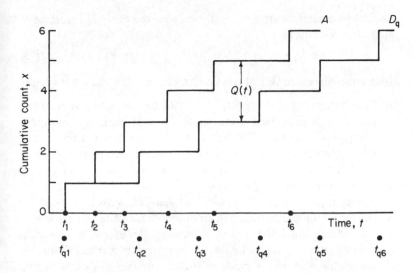

Figure 1.3 *Graphical representation of cumulative arrivals and departures from a queue*

For some purposes it is convenient to think of the graph in Fig. 1.3 as a graph of A versus t, i.e., the function $A(t)$, but for other purposes it is convenient to think of it as a graph of t versus A, i.e., the function $A^{-1}(x)$. If x is integer, we can consider it to be the label on the last arrival; for noninteger values we would interpret x as the cumulative number of arrivals or fractions thereof,

$$A^{-1}(x) = t_j \quad \text{for} \quad j-1 < x < j. \tag{1.2}$$

Perhaps it is better yet to consider this as simply a curve A in the t, x plane without specifying which variable is the 'independent' variable.

The second observer will record the times at which customers enter the server, along with the customer number assigned by the first observer. Let

t^*_{qj} = time customer number j leaves the queue and enters the service.

Whenever there is a queue of more than one customer, the order in which these customers enter the service need not be the same as the order in which they arrive. A rule describing how customers are selected from a queue is described in the queueing literature as 'queue discipline.' The discipline in which customers are served in order of their arrival is usually called 'first in, first out' or FIFO. This is the simplest to describe mathematically because the times t_{qj}^* must now satisfy the conditions

$$0 < t_{q1}^* \le t_{q2}^* \le t_{q3}^* , \ldots , \quad \text{for FIFO.} \tag{1.3}$$

Some common examples of queue disciplines other than FIFO are:

(a) Last in, first out (LIFO). Suppose that letters to be typed or order forms to be processed accumulate in a pile, each new addition being placed on top. The typist or clerk now services the letters (the customers) by taking each new task from the top of the pile. A newly arriving task will be the next to be served provided it can be served before another arrives.

(b) Service in random order (SIRO). Passengers waiting to board a bus might appear to board in an order which bears no relation to the order in which they arrive. Random order of service is usually defined to mean that whenever a customer is selected from the queue, the selection is made in such a way that any customer in the queue at the time of selection is equally likely to be chosen.

(c) Priority service. Particularly if one has not initially decomposed customers into categories and considered each category separately, one might order the customers in queue according to some identifiable characteristic (length of job or value). The next customer to enter the service is then the one in queue with the highest ranking (top priority) at the time. There are several variations on this depending upon whether or not a high priority customer must wait until the next service completion to enter the service or if it can displace a lower priority customer from the service.

For some purposes (particularly if one is interested only in counts of customers but not their identity) it may be convenient for the second observer simply to record the ordered times at which customers enter the service even though the queue discipline is not FIFO. He, in effect, relabels the customers and defines t_{qj} as the time of the jth departure from the queue so that

$$0 < t_{q1} \le t_{q2} \le t_{q3} \le , \ldots .$$

If one represents the times at which customers leave the queue by points on the real line as in Fig. 1.2(b), the set of times t_{qj} and t_{qj}^* represent simply two different labelings of the same set of points.

As with the times t_j, it is possible also to represent the t_{qj} by a graph

$$D_q(t) = \text{number of } t_{qj} \text{ with } t_{qj} \le t, \tag{1.4}$$

the cumulative number of departures from the queue by time t, or more generally

$$D_q(t) = \text{cumulative quantity or number to leave the queue by time } t.$$
(1.4a)

The inverse of this, $D_q^{-1}(x)$, describes the ordered departure times

$$D_q^{-1}(x) = t_{qj} \quad \text{for} \quad j-1 < x \leq j.$$
(1.5)

If one draws both $A(t)$ and $D_q(t)$ on the same graph, as in Fig. 1.3, the curves cannot cross because, for any t, the number of customers which have left cannot exceed the number which have arrived. The vertical distance between the two curves at any time, representing the number of customers who have arrived but have not yet left the queue, is

quantity or number in the queue (queue length)
$$= Q(t) = A(t) - D_q(t) \geq 0.$$
(1.6)

It is, of course, also true that
$$D_q^{-1}(x) - A^{-1}(x) \geq 0$$

because x customers cannot have left until at least x customers have arrived.

The curve $D_q^{(t)}$ does not display the queue discipline. It gives only the count of departures but not the identity, and its inverse gives only the ordered departure times t_{qj}. It is possible, however, to draw a graph D_q^* which does display both the departure times and the queue discipline. If we consider x as the independent variable, we can draw a function of x having values

$$t_{qj}^* \quad \text{for} \quad j-1 < x \leq j$$

instead of the t_{qj} which defined the curve D_q. Whereas the curve D_q described a monotone nondecreasing function of x or t, the curve D_j^* will not be monotone unless the customers are served in the order in which they arrive ($t_{qj} > t_{qk}$ if $j > k$), consequently the curve D_q^* will not generally define a single-valued function of t.

If we draw both A and D_q^* on the same graph as in Fig. 1.4, the horizontal distance from $t = 0$ to A at height $x, j-1 < x < j$ is the time t_j at which customer j arrives, and the horizontal distance to D_q^* is the time he left the queue. The difference between them, the horizontal distance from A to D_q^*, is the time which the jth customer spends in queue

$$w_j = t_{qj}^* - t_j \geq 0.$$
(1.7)

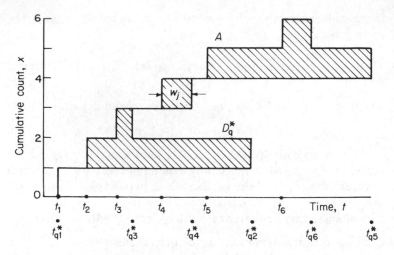

Figure 1.4 *Graphical representation of departure times*

This is also equal to the area of the rectangular strip between A and D_j^*, $i - 1 < x < j$.

Whereas Fig. 1.3 gives a simple geometric interpretation of $Q(t)$, Fig. 1.4 gives a simple geometric interpretation of the w_j. One could also identify the w_j from Fig. 1.3 if this graph were supplemented with some scheme for identifying which step in D_q gives the departure time associated with the jth step of A.

Since D_j^* shows both the departure times and the order of customers, it must also define $Q(t)$. The curves D_q^* and A enclose an area, the locus of all horizontal lines from A to D_q^*. If one draws a vertical line at time t, it will slice this area in such a way that any point x in this area is identified with a customer who has arrived but has not yet left. The total length of vertical line between A and D_q^* is the number in the queue $Q(t)$. This is also true of Fig. 1.3, but in Fig. 1.3 this is a single line segment.

If the queue discipline is FIFO the curves D_q and D_q^* are, of course, the same curves.

The above definitions of A, D_q, etc., are simply a description of what two observers recorded and involve no 'theory' of what happened to the customers in the queue. If we were to place a third observer downstream of the server, he could record similar information independent of what happens in the server. From this we can define t_{sj}^* as the time at which customer j leaves the service, t_{sj} as the ordered times at which customers leave, $D_s(t)$ as the cumulative number of customers to leave, and D_s^* as the curve defined by

$$t_{sj}^* \quad \text{for} \quad j-1 < x < j.$$

If we draw the curves A and D_s^* (instead of D_q^*) on the same graph, the horizontal distances would be interpreted as the times customers spend in the queue plus service (instead of just the queue) and a vertical slice of the area between A and D_s^* would determine the number of customers in the queue or in service (instead of $Q(t)$). Similarly, if we compare D_q^* and D_s^*, the horizontal distances would be the times customers spend in service and a vertical slice of the area between D_q^* and D_s^* would determine the number of customers in service at any time.

1.3 Modelling

In order to make predictions of what would happen in an experiment one has not done, one must relate the properties of D_q, D_s, etc., to any rules governing the behavior of the customers in the queue and the server. In a typical queueing problem one proposes one or more possible curves $A(t)$ or perhaps some probability distributions for $A(t)$, a description of the queue discipline (if relevant), and the manner in which the server operates. From this one wishes to evaluate, in effect, the curves D_q, D_s or any derived properties thereof, perhaps probability distributions of the w_j or $Q(t)$.

The rules governing the dynamics of the system could conceivably be quite complicated and involve interrelations between the service times, arrival times, customer types, etc., restricted only by the universal principle that a customer cannot leave before he has arrived or equivalently that the queue cannot be negative. Most systems which have been analyzed in the queueing literature, however, have rather simple rules. The mathematical complications are not directly associated with the queue dynamics, but with the stochastic analysis. Even though the postulated relations between arrival times and departure times appear quite simple, they lead to fairly complex relations between probability distributions for arrivals, departures, queue lengths, waits, etc.

A description of the server should at least define a relation between the curves D_q^* and D_s^*, i.e., between the t_{qj}^* and t_{sj}^*. The simplest rule is one in which the times each customer will be in service (or probability distributions for the service times) are given, i.e.,

$$s_j = t_{sj}^* - t_{qj}^* \qquad \text{for all} \quad j.$$

From this one can, of course, immediately construct the D_s^* from the D_q^*.

In some situations, for example, customers being served by taxis, the customer might consider his service to start when he boards the taxi and to be completed when he reaches his destination, so that the s_j would be his trip time. For the next customer who is waiting to be served, however, the relevant 'service time' is the time from the start of the last service until the taxi accepts him. Alternatively, he might consider his service to start when the taxi has discharged the previous customer and accepted his order and to end when he is discharged.

In the analysis of most queueing problems it is usually implied that the 'service time' is the time from the start of one service until the server is available to accept the next customer, since this is the time which is relevant to the evolution of the queue.

If one has m taxis serving the same queue, they could be serving as many as m customers simultaneously. A server of this type consisting of m separate servers, each of which serves only one customer at a time, is called an m-channel server. For such a system, a new customer can enter service as soon as any channel is free. The order in which customers complete service is not necessarily the same as the order in which they started service (i.e., the service discipline is not FIFO). A server which can, at the start of any service, accept several customers at the same time (such as a bus or elevator) is called a bulk server.

The rules governing the server will generally also specify that upon completion of one service at time t another service should start immediately provided that $Q(t) > 0$. If the server is a bulk server, the rules will also specify how many customers the server can accept, given the value of $Q(t)$. If $Q(t) = 0$, the rules should specify when the next service starts, usually when the next customer arrives.

For most service systems that one encounters in queueing applications it is quite easy to follow the rules iteratively (either numerically or graphically) and construct D_q^* and D_s^* from a given curve $A(t)$ and the service times (i.e., perform a 'simulation').

If, for example, the server is a single-channel server one would specify $A(t)$ or equivalently the t_j, the service times s_j, and the queue discipline. The iterative rules describing the evolution of the D_q and D_s are that

$$t_{sj}^* = t_{qj}^* + s_j;$$

and, if the queue discipline is FIFO, the $j + 1$th customer starts service at time t_{sj}^* if he has arrived, i.e. if $t_{j+1} < t_{sj}^*$, but, if not, he enters service as soon as he does arrive at time t_{j+1}. Thus, $t_{qj}^* = t_{qj}$, $t_{sj}^* = t_{sj}$ and

$$t_{qj+1} = \max(t_{j+1}, t_{sj}) = \max(t_{j+1}, t_{qj} + s_j). \tag{1.8}$$

Starting from an initial condition that the system is empty and $t_{q1} = t_1$, (1.8) determines each t_{qj+1} from the previous one, t_{qj}.

If we subtract t_{j+1} from both sides of (1.8) we can also write it in the form

$$t_{qj+1} - t_{j+1} = \max(0, t_{qj} - t_{j+1} + s_j). \tag{1.9}$$

For FIFO queue discipline

$$w_j = t_{qj} - t_j,$$

therefore, the w_j satisfy the equations

$$w_{j+1} = \max(0, w_j + s_j - (t_{j+1} - t_j)). \tag{1.10}$$

This describes the waiting times iteratively in terms of the service times and interarrival times $t_{j+1} - t_j$.

1.4 Averages

For most queueing systems, it requires only elementary mathematics to describe the detailed dynamics of the system. It is, essentially, the approximate theory of queues which is complicated. In analyzing the behavior of queues, one does not care to observe the arrival and departure times of every customer. On the one hand, this involves tedious manipulations. On the other hand, they are not very interesting data because many of them could not be reproduced if the experiment were repeated. One would prefer to specify only a few characteristics such as some average arrival rate or service rate, things which are nearly reproducible.

Even if one could describe in detail exactly how large the queue would be at every instant of time, one would probably disregard much of the detail. One would prefer to have only some approximate description or measure of performance. This is, in essence, why one treats queueing phenomena as stochastic processes. One only wishes to consider the average behavior of the system over a range of conditions, not the details of what happens in any particular experiment.

Whether one treats the system stochastically or deterministically, there are certain gross properties one may wish to calculate; for example, the average wait in queue for a set of n customers or the time average queue length over some period of time.

The average time in queue for customers $j+1$ to $j+n$ inclusive is defined as

$$\langle w_k \rangle = \frac{1}{n} \sum_{k=j+1}^{j+n} w_k = \frac{1}{n} \sum_{k=j+1}^{j+n} (t_{qk}^* - t_k). \qquad (1.11)$$

The w_k can also be interpreted as the area of a horizontal strip $k-1 < x < k$ between A and D_q^*. The sum of the w_k is, therefore, the area enclosed by A, D_q^* and two horizontal lines $x = j$ and $x = j+n$ as shown in Fig. 1.5(a).

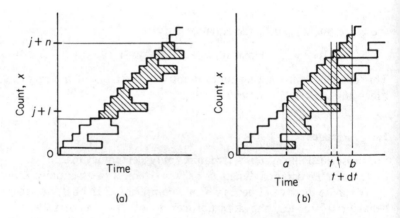

Figure 1.5   *Areas defining average wait and queue length*

The average queue length during some time interval (a, b) is defined as

$$\langle Q(t) \rangle = \frac{1}{(b-a)} \int_a^b Q(t)\, dt. \qquad (1.12)$$

Since $Q(t)$ can be interpreted as the length of cross section cut from the region between A and D_q^* by a vertical line at t, $Q(t)\,dt$ represents the area cut from this region by a vertical strip between t and $t + dt$. Thus the integral from a to b represents the area enclosed by A, D_q^* and two vertical lines at $t = a$ and $t = b$, as in Fig. 1.5(b).

If, in (1.12), we chose a and b as any times for which $Q(a) = Q(b) = 0$, and in (1.11) we chose $j = A(a), j+n = A(b)$, then the two areas in Fig. 1.5(a) and (b) would be the same areas. Both the horizontal and vertical lines of Fig. 1.5(a) and (b) would cut the region between A and D_q^* at a single point. The total queueing time during the time interval (a, b) would be the same as the total queueing time for customers $j + 1$ to $j + n$. In (1.11) this is represented as the sum of horizontal strips whereas in (1.12) it is represented as the sum of vertical strips. With the a, b, j, and n chosen in this way, it follows that

$$(b - a)\langle Q(t)\rangle = n\langle w_j\rangle = \text{total queueing time}$$

or

$$\langle Q(t)\rangle = \frac{n}{(b-a)}\langle w_j\rangle. \tag{1.13}$$

We can also define

$$\lambda_{ab} = n/(b - a) \tag{1.14}$$

as the average arrival rate during the time interval (a, b).

If the queue behavior is such that the queue vanishes repeatedly, every day at midnight or at other perhaps irregular (maybe stochastic) times with a finite spacing, the above relations would be valid for any or all choices of times a and b when the queue vanishes (not just consecutive times). Even if the queue does not vanish at times a and b, but it vanishes at many other times between a and b including some times close to a and b, the areas in (1.11) and (1.12) would differ only at the ends of the region between A and D_q^* and from a or b to the nearest time where $Q(t)$ does vanish. If these end areas are negligible compared with the total areas (1.13) is still approximately correct.

Although (1.13) is exactly correct if $Q(a) = Q(b) = 0$, essentially as a direct consequence of the definitions of the averages, and it may be approximately true under more general conditions, the $\langle w_j\rangle$, $\langle Q(t)\rangle$, and λ_{ab} will generally depend upon the particular choice of a and b.

Much of the mathematical literature on queueing theory deals with what is known as 'stationary arrivals' generated by a hypothetical source which operates from $t = -\infty$ to $t = +\infty$. The arrivals are assumed to have certain stochastic properties but of such a nature that for $a \to -\infty$ and $b \to +\infty$ each of the above averages, particularly the λ_{ab}, has a well defined limit. If such limits exist, (1.13) must, of course, be valid for $a \to -\infty$ and $b \to +\infty$. Theorems relating to the validity of (1.13) can be quite sophisticated,[†] but the complications are

† Little, J. D. C. (1961) A proof for queueing formula $L = \lambda W$. *Operations Research*, **9**, 383–7.
Jewell, W. S. (1967) A simple proof of: $L = \lambda W$. *Operations Research*, **15**, 1109–16.

mainly associated with the mathematical conditions which will guarantee the existence of the appropriate limits. From the point of view of practical applications, this is, however, rather academic since no real process runs from $t = -\infty$ to $t = +\infty$.

Equation (1.13) has an obvious generalization to cases in which the x coordinate measures some substance other than numbers of customers. Suppose, for example, that $A(t)$ measured the cumulative arrivals of a substance which arrives in discrete units of size a_k at times t_k, and leaves in the same units of size a_k at times t_{qk}^*. If $Q(t)$ now measures the amount of substance in the reservoir and $Q(a) = Q(b) = 0$, the area between A and D_q^* from a to b would be

$$\int_a^a Q(t)\,\mathrm{d}t = \sum_{k=j+1}^{j+n} a_k(t_{qk}^* - t_k),$$

with the sum extending over all arrivals between times a and b, or

$$\langle Q(t) \rangle = \frac{1}{(b-a)} \sum_{k=j+1}^{j+n} a_k(t_{qk}^* - t_k). \tag{1.15}$$

There are (at least) two possible interpretations of the right hand side of (1.15). If a_k represents the value of the kth arriving object or the cost per unit time for delay of the object (interest cost on the value), we could interpret

$$\frac{1}{n} \sum_{k=j}^{j+1} a_k(t_{qk}^* - t_k)$$

as the average cost of delay per arrival and write (1.15) as

$$\langle Q(t) \rangle = \lambda_{ab} \left[\frac{1}{n} \sum_{k=j}^{j+n} a_k(t_{qk}^* - t_k) \right] \tag{1.16}$$

with λ_{ab} the (average) arrival rate of objects. Alternatively we might interpret

$$\frac{1}{\displaystyle\sum_{k=j}^{j+n} a_k} \sum_{k=j}^{j+n} a_k(t_{qk} - t_k)$$

as the delay per unit of substance (for example, delay per person if the arrivals are buses with a_k passengers per bus) and write (1.15) as

$$\langle Q(t) \rangle = \left[\frac{\sum\limits_{k=j}^{j+n} a_k}{(b-a)} \right] \left[\frac{1}{\sum\limits_{k=j}^{j+n} a_k} \sum_{k=j}^{j+n} a_k (t_{qk}^* - t_k) \right] \qquad (1.17)$$

$$= \text{(arrival rate of substance)} \times \text{(average delay per unit of substance)}.$$

The above formulas were derived from the geometric properties of two curves (A and D_q^*) but did not depend upon how the curves were generated. If the curves had been A and D_s^* instead of A and D_q^*, and the system was empty at times a and b, we would interpret (1.13) or (1.17) as

$$\langle Q_s(t) \rangle = \text{average substance in the system}$$
$$= \text{(arrival rate of substance)} \times \text{(average time in system per unit of substance)}. \qquad (1.18)$$

Similarly, if the curves were D_q^* and D_s^* but there was nothing in the server at times a and b, (1.13) or (1.17) could be written as

$$\text{average substance in service} = \text{(arrival rate of substance)}$$
$$\times \text{(average time in service per unit of substance)}. \qquad (1.19)$$

Furthermore, if a and b were times when the system is empty, the arrival rates in (1.17), (1.18), and (1.19) would be all the same and the average delay times refer to the same set of objects.

The above formulas are true regardless of the queue or server discipline, but the average queue length, which does not recognize the identity of customers, could have been evaluated from the curves A and D_q. If a change in the queue discipline does not change the curve D_q, i.e., the times at which customers enter the server, then the average wait per customer is independent of the queue discipline. The D_q will be independent of queue discipline, thus unaffected by an interchange of customers, if and only if the service times of all customers are equal (or if the service times are random, they are 'interchangeable').

If this is true, the advantage of FIFO discipline over other types of queue disciplines is related not to the average delay but to the variations in delay about the average. Obviously, last in, first out discipline gives a high proportion of very short delays (less than one service time) but also some very long delays.

1.5 Applications of $L = \lambda W$

In a typical queueing problem, one specifies the arrival rate of customers, λ_{ab}, and the service times. One wishes to evaluate (among other things) the average queue length $\langle Q(t) \rangle$ and/or the average delay per customer $\langle w_j \rangle$. Equation (1.13) relates these two unknowns in a simple way, so it suffices to determine either one or the other.

Equation (1.19), however, has some more direct applications because both factors on the right hand side are usually given. Consequently, one can immediately evaluate the average number of customers in service, provided, of course, that the long time arrival rate of customers is sufficiently low that the system will empty occasionally, i.e., the server is capable of serving customers fast enough eventually to keep up with the arrivals.

If the server is an m-channel server, with each channel serving at most one customer at a time, the number of customers in service is the same as the number of busy servers. Thus (1.19) determines also the (time) average number of servers that are busy, which is obviously a lower bound on the number of servers m which one needs to keep up with the arrivals. A telephone company, for example, might know the frequency of calls (arrival rate) between two cities and the average duration of a call (service time). It wishes to know the minimum number of channels it must provide to handle the traffic. An airport designer may know how many aircraft arrivals are expected each day and the average 'turn around time', i.e., the average time an aircraft occupies a gate position (service time). He wishes to know the minimum number of gate positions he must build.

In each case, the designer would also like to know something about the peaking of demand and the delays that would result from various choices of m, but (1.19) will not determine that. In fact, if all channels are identical, presumably the service time of a customer will be independent of which channel it uses and, consequently, also independent of the number of channels. Thus the right hand side of (1.19) does not depend upon m (provided m is large enough eventually to serve the arrivals) and the average number of busy servers is independent of the number of channels.

The difference between a service with many channels and one with only the minimum number is that the former can serve customers with less delay in queue. If one has an arbitrarily large number of channels, many channels will be used during temporary surges in the arrivals but during lulls relatively few will be used. If one has only the minimum number of servers, a queue forms during the surges but the customers in queue are served during the lulls; the servers are kept busy all of the time serving either new arrivals or those in queue. The time average number of busy servers is the same in both cases, but the former has larger fluctuation.

Equation (1.19) also gives some interesting information for a single-channel server. Again one typically knows the arrival rate and average service time, so (1.19) determines the time average number of customers in service. For a single-channel server, however, the number in service at any time can be only 0 or 1. The time average number in service is, therefore, the same as the fraction of time the one server is busy. The right hand side of (1.19), $\lambda_{ab} \langle s_j \rangle$, is, in this case, called the 'traffic intensity' usually denoted by ρ:

$$0 \le \rho = \lambda_{ab} \langle s_j \rangle \le 1. \tag{1.20}$$

The quantity $1 - \rho$ is the fraction of time the server is idle. If a customer arrives at a random time uniformly distributed over the time interval (a, b), $1 - \rho$ can also be interpreted as the probability that the customer finds the server idle and can enter service with no delay.

1.6 Other graphical representations

In Section 1.2 we represented $A(t)$ and $D_q(t)$ as two curves on the same graph, i.e., as the locus of points in the (x, t) plane with coordinates $(A(t), t)$ and $(D_q(t), t)$ respectively, or equivalently $(x, A^{-1}(x))$ and $(x, D_q^{-1}(x))$. This is the most common way of representing these quantities graphically because it shows very conveniently most of the quantities one wishes to observe, particularly if the queue discipline is FIFO and $D_q^* = D_q$. The advantage of this type of graphical representation of the data t_j, t_{qj}, etc., over other possible schemes derives from the fact that one can easily visualize the geometrical addition or subtraction of line segments or areas. The graphs conveniently show geometrically the subtraction $A(t) - D_q(t)$ to give $Q(t)$ and, for FIFO queue discipline, the geometric subtraction of line segments in a different direction $D_q^{-1}(x) - A^{-1}(x)$ to give the wait $w(x)$. Furthermore, it conveniently shows addition of areas to give the total waiting time. If these are the features of primary interest, it is difficult to imagine how one could find any better way to show all of these on the same graph.

If one wishes to compare the behavior of the curves $A^{(1)}(t)$, $D_q^{(1)}(t)$ as observed on one day with a new pair of curves $A^{(2)}(t)$, $D_q^{(2)}(t)$ observed on another day, the fact that one must draw two curves for each day and then compare the *pairs* of curves may be awkward. It is possible, however, to show the evolution of both $A(t)$ and $D_q(t)$ simultaneously by a single curve if one goes to a three-dimensional space. In an (x, y, t) space one can draw a curve $(A(t), D_q(t), t)$ or in a (t, t_q, x) space one can draw a curve $(A^{-1}(x), D_q^{-1}(x), x)$.

Each of these is a step function curve. The former moves a unit step in the x direction at each time t_j and in the y direction at each time t_{qj}; whereas the latter curve jumps from (t, t_q) coordinates (t_j, t_{qj}) to $(t_{j+1}, t_{q, j+1})$ when x passes j. The two curves $(A(t), t)$ and $(D_q(t), t)$ are the projections of $(A(t), D_q(t), t)$ onto the (x, t) and (y, t) planes, respectively. Correspondingly, the curves $(A^{-1}(x), x)$ and $(D_q^{-1}(x), x)$ are the projections of $(A^{-1}(x), D_q^{-1}(x), x)$ onto the (t, x) and (t_q, x) planes, respectively. We have, of course, previously drawn these projections on the same graph. We could have drawn them on separate graphs but would then lose the simple geometric interpretation of queue length and wait in queue which resulted from subtracting distances between the curves A and D_q.

There is still a third projection of each of these three-dimensional curves, the projections on the (x, y) or (t, t_q) planes. These are curves having a parametric representation $(A(t), D_q(t))$ and $(A^{-1}(x), D_q^{-1}(x))$, respectively, as shown in Fig. 1.6. Actually, the latter 'curve' is only a sequence of points since it moves only at integer x, but we can arbitrarily join these points by a piecewise linear curve. One may also label the time t or count x as a parameter along the curve, particularly since the only relevant parameter values are the discrete times t_j and t_{qj} or the integer values of x.

On the $(A(t), D_q(t))$ curve, a horizontal step of the curve has a length equal to the number of successive arrivals before a departure, and a vertical step has a length equal to the number of successive departures before the next arrival. If a t_j should be equal to some t_{qk}, we would have simultaneous horizontal and vertical steps. This would certainly occur at times when a customer arrives and finds the server empty, since then $t_j = t_{qj}$. It seems most natural to represent simultaneous arrivals and departures by a single line of slope 1.

That $A(t) \geq D_q(t)$ and $D_q^{-1}(x) \geq A^{-1}(x)$ was displayed in Fig. 1.3 by the D_q curve always being below or to the right of A. One needed to

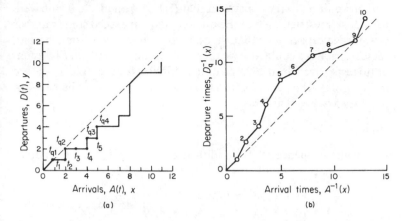

Figure 1.6 *Parametric representations of* $D(t)$ *versus* $A(t)$ *and* $D_q^{-1}(x)$ *versus* $A^{-1}(x)$

compare two curves to see this. In Fig. 1.6, however, this important property is shown by the one curve $(A(t), D_q(t))$ which must lie always on or below the line $x = y$, or by $(A^{-1}(x), D_q^{-1}(x))$ which must lie on or above the line $t = t_q$. One is still comparing two curves, but the straight line does not depend upon the evolution of the system. If one superimposes curves obtained on several days, the boundary line is the same for all days.

A graph such as Fig. 1.6 displays quite different aspects of the system behavior from Fig. 1.3. The former shows in a more convenient way comparative properties of the arrival and departure times (for FIFO), but it does not show conveniently any comparative properties of counts and times because one or the other is represented only as a parameter along the curve. Fig. 1.6(a) does not show waiting times conveniently because these involve the time parameter t; Fig. 1.6(b) does not show queue lengths conveniently because these involve the customer count parameter x. Neither graph shows the arrival rate, departure rate. $\langle w_j \rangle$, or $\langle Q(t) \rangle$ since these all involve both counts and time.

Fig. 1.6(a) does show the queue length identified with any point on the curve having integer coordinates; it is either the horizontal or vertical distance from the curve to the 45° line. Similarly, Fig. 1.6(b) shows the waiting time associated with any point (t_j, t_{qj}) as the horizontal or vertical distance to the 45° line.

If, in addition to the constraint that $Q(t) \geq 0$ and $w_j \geq 0$, one were to impose a restriction, that the queue could not exceed some number c (a storage capacity) or that the wait could not exceed some bound, these restrictions could also be represented conveniently in Fig. 1.6(a) or (b) respectively by drawing another $45°$ line; for example, $x - y = c$ in Fig. 1.6(a). The curve $(A(t), D_q(t))$ would then always lie between the lines $x - y = 0$ and $x - y = c$.

Problems

1.1 From a sequence of 150 random digits $X_i = 0, 1, \ldots, 9$, generate a sequence of numbers

$$Y_i = \begin{cases} 1 & \text{if} \quad X_i = 0, 1, 2, \text{ or } 3 \\ 0 & \text{if} \quad X_i = 4, 5, 6, 7, 8, \text{ or } 9 \end{cases} \quad i = 1, \ldots, 150.$$

From a different sequence of 150 random digits $X_i' = 0, 1, \ldots, 9$ generate a sequence

$$Y_i' = \begin{cases} 1 & \text{if} \quad X_i' = 0, 1, 2, 3 \text{ or } 4 \\ 0 & \text{if} \quad X_i' = 5, 6, 7, 8, \text{ or } 9 \end{cases} \quad i = 1, \ldots, 150.$$

The X_i, X_i' may be obtained from tables of random digits or one may use last digits of consecutive telephone numbers from a telephone book (excluding any numbers which may be listed twice, because a person may have both a business and personal listing, for example).

Consider a queueing system for which customers arrive and leave only at integer values of time $t = i$. One customer arrives at time i if $Y_i = 1$, none if $Y_i = 0$. If a customer is in service at time $i - 1$, it will leave the service at time i if $Y_i' = 1$, otherwise no customer leaves at time i. The server serves only one customer at a time.

Draw graphs of $A(t)$, $D_s(t)$, and $Q_s(t) = A(t) - D_s(t)$ on 10 squares to the inch graph paper with a scale of 20 time units to the inch and 10 customers to the inch. Also draw graphs of $(A(t), D_s(t))$ and $(A^{-1}(x), D_s^{-1}(x))$. Interchange the sequences Y_i and Y_i' and draw the corresponding graphs.

Note: This may be assigned as a class exercise. Each student should select a different set of random digits. One can then imagine that the different curves represent the results of some experiment which was repeated many times. The class should then compare the graphs obtained by different students.

1.2 If on a graph of $Q(t)$ versus t one sees that an arriving customer causes $Q(t)$ to increase from k to $k+1$, how would one identify when he left the queue if the queue discipline is last in, first out? How would one identify the same thing from graphs of $A(t)$ and $D_q(t)$?

1.3 (a) Let $0 = t_1 < t_2 < \ldots t_n$ be ordered arrival times and $0 < t_{q1} < t_{q2} < \ldots t_{qn} = t_n$ ordered departure times from a queue which vanishes at time 0 and t_n. If $t^*_{qj} = t_{qn_j}$ is the departure time of customer j, show that the sum of the squares of the delays

$$\sum_{j=1}^{n} (t_{qn_j} - t_j)^2$$

is least if $n_j = j$, i.e., for FIFO service. The t_{qj} are assumed to be independent of the order of service.

(b) As a generalization of (a), suppose the cost of dealy to the $j$the customer is a function $P(w_j)$ of the delay $w_j = t_{qn_j} - t_j$ with the function $P(x)$ the same for all customers. The total cost of delay to all customers is

$$\sum_{j=1}^{n} P(w_j).$$

If the marginal cost per unit delay $p(x)$,

$$p(x) = dP(x)/dx, \quad P(x) = \int_0^x p(x')\,dx',$$

is a monotone increasing function of x, show that the total cost of delay is least for FIFO.

CHAPTER 2

Deterministic fluid
approximation – single server

2.1 Introduction

To analyze the behavior of some existing service facility which serves a
single category of customers, imagine that one were to record the
arrival and departure times of all customers over some very long
period of time, possibly several years. Such data do exist for some real
systems. For example, any computer controlled traffic signal system is
connected with many permanently installed vehicle detectors. At any
particular traffic signal (server) there is likely to be one or more
detectors upstream of the signal and also detectors downstream
(perhaps near the next downstream signal). These detectors transmit
an electrical pulse to the computer every time a vehicle passes. Most of
these data are discarded after use, but they could be kept on a
magnetic tape. Any airport also maintains a record of all aircraft
movements including arrivals and departures from the runway and
gate positions.

In the last chapter we discussed some of the microscopic properties
of the $A(t)$, $D_q(t)$ curves; that they have integer steps at each arrival
time. If one were to draw a graph of $A(t)$ and $D_q(t)$ for several years on
a scale such that one could see each arrival, it might require a square
mile of graph paper. If, however, one were to rescale the graph, one
would see different types of time-dependent phenomena depending
upon the scales of counts and time.

If you were counting cars on a highway, for example, which at
various times might carry flows of the order of 1000 cars per hour, one
could see the integer steps in $A(t)$ if the scales of time and count were
comparable with, say, 10 seconds to the cm and 3 counts per cm. If,
however, one were to choose a scale of about 1 min per cm and 20
counts per cm, the integer steps would be too fine to show very clearly.
The $A(t)$ and $D_q(t)$ curves would likely appear as a nearly smooth

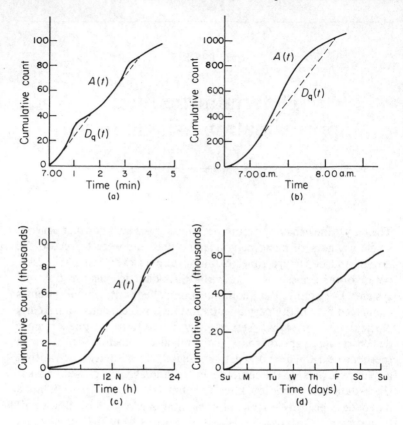

Figure 2.1 *Cumulative arrivals on various time scales*

curve, as illustrated in Fig. 2.1(a), having some wiggles of rather erratic form caused by 'random surges'.

Depending upon how steep the curves become as a result of the wiggles and how fast the server operates, the surges may or may not cause a queue. If the server has nearly equal service times for all cutomers, the $D_q(t)$ curve might look like the broken line curve of Fig. 2.1(a), which shows a queue whenever the surges generate a slope temporarily exceeding the service rate.

If we draw the curve on a still coarser time scale of say 20 minutes per cm and 200 counts per cm, the wiggles in the $A(t)$ might not show very clearly, but one would see the 'peak demands' or 'rush hours'. For example, between 7.00 a.m. and 8.00 a.m. of some day, the curve for

$A(t)$ might be of the type shown in Fig. 2.1(b). The $A(t)$ curve may be quite smooth, but not linear. If there is a portion of the curve where the arrival rate exceeds the service rate, the $D_q(t)$ could deviate appreciably from $A(t)$ showing queues possibly of the order of 100 cars.

On a scale of a few hours per cm and several thousand counts per cm, the difference between $A(t)$ and $D_q(t)$ would likely be too small to measure (the queues are not measured in thousands), but over a reasonable width of paper one sees the 24-hour flow pattern. It might show a morning and evening peak but a very low arrival rate between 2.00 a.m. and 5.00 a.m. with no queue (even on the scale of Fig. 2.1(a)) as in Fig. 2.1(c).

If the graph is continued for a week as in Fig. 2.1(d), we would probably see different patterns on different days, particularly Sunday, Monday, Friday, and Saturday. If it is drawn on a scale of a week per cm, we would no longer see the daily rush hours. We might see some variation in the daily counts within the week (particularly the weekend) but over many weeks there is likely also to be some seasonal variation. Finally, if one draws the graph on a scale of several months per cm, one would no longer see the daily variations. The seasonal variations would still be visible, but, in addition, one might see a gradual growth of traffic from one year to the next.

To analyze queueing delays, it is obviously not very helpful to draw $A(t)$ and $D_q(t)$ on such a scale that one cannot measure the difference between the curves. One would not ordinarily draw a graph on a scale such as Fig. 2.1(d) (or a coarser scale), but would observe that the queue (almost always) vanishes around 2.00 a.m. to 5.00 a.m. Instead of continuing the curve $A(t)$ for several days, one could reset the counters to zero at the same time each day and draw separate curves $A^{(j)}(t)$ and $D_q^{(j)}(t)$ for each jth day, i.e., a set of curves of the type shown in Fig. 2.1(c).

If we compare the curves on various days, we should, of course, find that some days have similar patterns (successive Fridays, for example). We would first try to classify the days in some systematic way so that all curves in the same class can be compared as being nearly 'equivalent'. No matter how one does this, however, there are likely to be some curves which are unlike any others. There are days when there were failures in the service, unusual patterns caused by a snow storm or whatever, which occur only once or twice in several years and possibly never in quite the same way again.

In the analysis of most practical problems, one would probably not try to analyze the behavior of the system for all time periods. Presumably, there are certain things about the system behavior which are undesirable and which one wishes to correct through some modification in design, strategy of operation or whatever. Since it is expensive to collect and analyze data, one must first decide 'what was the problem?' or 'what are the most important of several possible problems?'

The type of techniques one will use to analyze the system depends upon whether one wishes to investigate the queueing due to the wiggles in $A(t)$ as illustrated in Fig. 2.1(a), the rush hours as in Fig. 2.1(b), or the unusual events. If it is the rush hour, one must further decide if the problem is primarily the weekday rush, the Monday–Friday rush or the weekend demands (as for recreational facilities).

To design a facility to accommodate the unusual events often means merely that one has some emergency procedures for diverting the customers (aircraft are sent to another airport in a snow storm or an announcement is made that some facility will be closed). The evaluation does not typically involve conventional queueing methods because the inconvenience may not be the usual delay associated with customers waiting to be served. The inconvenience may be difficult to quantify, yet many facilities (particularly public facilities) are designed in response to complaints about service during unusual situations. If the performance of a system is mainly a 'political' issue, there is no point in making an economic evaluation of delays during typical days. (People may want a rapid transit system no matter what it costs.)

We will be concerned here, by implication at least, primarily with patterns of system behavior that recur many times or are (partially) predictable. If it costs more to build a facility with a larger service rate, an efficiently designed system will always cause some delays, because there is no benefit associated with any excess capacity which is never used. The benefit associated with an increment of capacity that is used for only a short period of time or infrequently is also very small. Consequently, the capacity should always be somewhat less than the maximum demand during some time period. In principle, the proper choice of capacity should involve a compromise between the cost of a large facility and the inconvenience of delays for a smaller facility.

Even after one has classified the days into well defined categories, one will still find that the $A^{(j)}(t)$ curves within the same class are different in various ways. The arrival times of the customers are not likely to be exactly the same. Furthermore, although the curves on different days may have nearly the same form when drawn on a scale such as Fig. 2.1(b) or (c), the wiggles on a scale corresponding to Fig. 2.1(a) do not occur at the same times or have the same shapes.

Most of the literature on queueing theory deals with the analysis of queueing on a scale corresponding to Fig. 2.1(a). It is generally assumed that if one takes the arithmetic average of the $A^{(j)}(t)$ over many days,

$$\langle A^{(j)}(t) \rangle = \frac{1}{n} \sum_{j=1}^{n} A^{(j)}(t), \qquad (2.1)$$

that this averaging will smooth out most of the wiggles of the individual curves (for sufficiently large n) giving a curve for $\langle A(t) \rangle$ that is nearly linear over some appropriate time interval. Actually most of the theory deals with 'stationary processes' corresponding to some hypothetical process that runs from $t = -\infty$ to $t = +\infty$. It is usually further assumed that the arrival rate, i.e., the slope of the linear $\langle A^{(j)}(t) \rangle$, is less than the (average) service rate while the server is busy.

Whereas $Q^{(j)}(t) \geq 0$ on every day and, consequently, the average of $Q^{(j)}(t)$ over n days is positive, if one had a hypothetical process which had an arrival curve $\langle A^{(j)}(t) \rangle$ and a server which, while busy, serves at a constant average rate larger than the arrival rate, there would be no queue. Thus, the average $\langle Q^{(j)}(t) \rangle$ is not the same as the queue generated by the average arrival curve. The former is always larger than the latter because the queues generated by random surges are not compensated by a negative queue during the lulls. Despite the fact that the $Q^{(j)}(t)$ are not the same on different days, one will likely find that the total delay over some sufficiently long time (or the time average of $Q^{(j)}(t)$ or the average wait of many customers on the jth day) is nearly the same on all days.

In practical applications, however, many (perhaps most) queueing problems are of a type analogous to that shown in Fig. 2.1(b). If one compares these curves on different days, one still sees that the curves have wiggles in different places on different days but the amplitude of the wiggles is small compared with a typical queue length during the rush. Except for effects caused by wiggles near the beginning or end of the queueing period, the $\langle Q^{(j)}(t) \rangle$ is nearly the same as would be generated by a hypothetical arrival process $\langle A^{(j)}(t) \rangle$ served by an average server.

In analyzing any queueing problems in which queues form systematically at nearly the same time on all days (of the same category) it is convenient artificially to separate the queue length into two parts, a part which would be generated by a hypothetical arrival process $\langle A^{(j)}(t)\rangle$ served by an average server, and the excess queue caused by the daily variations about these averages. The former part is called the 'deterministic queue'. If the latter can be described by some probability model, it will be called the 'stochastic queue'.

In the following, we will first consider a variety of problems which can be analyzed approximately from the evaluation of the deterministic queues. We will later describe some of the stochastic effects associated with some relatively simple types of systems.

In addition to the examples described in Fig. 2.1 primarily for a system having predictable rush hours but a steady server, deterministic approximations are also useful for describing any system for which a queue forms for predictable reasons. There are many systems in which the service is interrupted for specified times long enough for a sizeable queue to form. The service may be interrupted for one class of customers because the server is being used to serve another class of customers or perform some other function. At a highway traffic signal, for example, the signal turns red while a traffic intersection is used to serve another traffic stream. A bus will interrupt the loading of passengers who have waited for the bus because the bus must be used to transport the passengers somewhere. A server may be interrupted also for repairs.

The objective in modeling any such system is, of course, to relate the behavior of the system to various parameters associated with the arrivals or the server so that one can predict how some hypothetical system with different parameters will perform.

2.2 A rush hour

To analyze a rush hour of the type shown in Fig. 2.1(b) and to estimate what queues would exist for various service rates, one must first collect data to estimate or predict a possible arrival curve $A(t)$ or an average curve $\langle A^{(j)}(t)\rangle$.

In practical applications, as for example in counting cars on a highway where there is no automatic recording equipment, the data which are often recorded consist only of the counts of arrivals during consecutive time intervals of, say, 5, 10, or 15 minutes. These data are also commonly displayed as a histogram of counts as in Fig. 2.2 which one interprets as some step function approximation to a smooth function $\lambda(t)$. These observations might not be repeated on another day, it being assumed that the results would be reproducible within the typical range of statistical fluctuations, and that the hypothetical

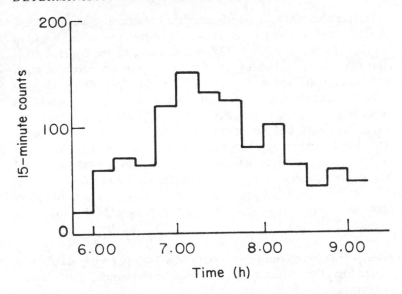

Figure 2.2   *Histogram of 15-minute counts*

$\lambda(t)$ which one would presumably obtain by averaging over many days is some suitable smoothing of the histogram.

From the histogram one can, of course, evaluate $A(t)$ at the discrete times corresponding to the ends of the counting interval. Lacking a statistical model, one would probably assume that the counts within the time intervals were nearly uniformly distributed over the interval, that the $A(t)$ therefore increases nearly linearly between the ends of the time intervals, that the $\langle A^{(j)}(t) \rangle$ would be a smoothing of $A(t)$ on some appropriate time scale, and that $\lambda(t)$ is some suitable smoothing of the histogram.

Since the questions we will be asking relate directly to the curve $A(t)$ and areas between it and some proposed $D_q(t)$, it is more important that one has an accurate estimate of $A(t)$ (for all t) than $\lambda(t)$. If one is smoothing 'by eye', it is generally better to smooth the $A(t)$ directly and evaluate $\lambda(t)$, if relevant, as the derivative of the smoothed $A(t)$ than to estimate $\lambda(t)$ from a smoothing of the histogram and evaluate $A(t)$ by integrating the smoothed $\lambda(t)$. In doing the latter, one might, in an attempt to follow the histogram 'locally', make systematic errors which would accumulate in the integration to obtain $A(t)$, and therefore cause inaccurate estimates of the $Q(t)$.

Having obtained an approximate deterministic $A(t)$, we might now imagine that we have a server which operates at some constant average rate μ when busy. The construction of the curve $D_q(t)$ is illustrated in Fig. 2.3(a) for a typical rush hour specified by a given arrival curve $A(t)$ and service rate μ. The curve $D_q(t)$ follows $A(t)$ very closely (essentially zero queue) until a time to when the arrival rate $\lambda(t)$ is equal to μ. If for some range of t with $t > t_0$, $\lambda(t) > \mu$, a queue starts to form at time t_0 and the service rate remains constant at the value μ. Thus the departure curve becomes a straight line of slope μ tangent to $A(t)$ at t_0 and extending from t_0 until some time t_3 when the arrival rate has been below μ long enough for the service to have caught up with the arrivals, i.e., $D_q(t_3) = A(t_3)$. After time t_3, the queue stays zero, i.e., $D_q(t) = A(t)$ for $t > t_3$ until such time as the $\lambda(t)$ again exceeds μ.

Note that the graphical construction of $D_q(t)$ is very easy for any given μ. One merely pushes a straight edge at slope μ against the curve $A(t)$ to form the tangent at t_0 and draws the line from t_0 until it meets $A(t)$ again.

One can see immediately from Fig. 2.3(a) that, for any smooth $A(t)$, $Q(t)$ grows quadratically in time from time t_0 but vanishes linearly in t near t_3.

It is often convenient also to draw graphs of $\lambda(t)$ and the actual departure rate $\mu(t)$ as in Fig. 2.3(b). Until time t_0, $\mu(t) = \lambda(t) \leq \mu$. For some time after time t_0, however, $\mu(t) = \mu \leq \lambda(t)$ as shown by the broken line. If $\lambda(t)$ reaches a maximum at time t_1 and decreases, it will equal μ again at some time t_2.

The length of the queue at time t

$$Q(t) = A(t) - D_q(t) = \int_{t_0}^{t} [\lambda(\tau) - \mu(\tau)] \, d\tau \qquad (2.2)$$

is represented in Fig. 2.3(b) by the shaded area between $\mu(\tau)$ and $\lambda(\tau)$. It reaches a maximum at time t_2 when $\lambda(\tau) = \mu$. After time t_2, the queue decreases until time t_3 when the area between $\lambda(t)$ and μ of Fig. 2.3(b) from time t_2 to t_3 is equal to the corresponding area between times t_1 and t_2. Note that $\mu(t)$ will, generally, have a discontinuity at time t_3 when, as the queue vanishes, $\mu(t)$ drops from μ to $\lambda(t_3)$.

Although Fig. 2.3(b) shows clearly the evolution of the $\lambda(t)$ and $\mu(t)$, it does not show conveniently such things as waits, queue lengths, etc., as does Fig. 2.3(a).

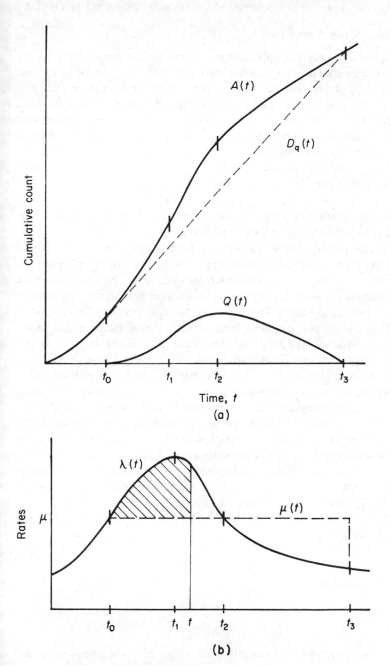

Figure 2.3 *Graphical construction of queue evolution*

2.3 A slight overload

In most engineering applications, the evaluation of delays is only the first step in an analysis, the final result of which is a decision as to what to build. If costs of delays are large (in some sense) relative to cost of construction, one should build a facility with a μ very close to the peak arrival rate, thereby keeping the delays low. If, however, construction costs are high, one builds a facility only large enough to serve all customers eventually, but certainly not to serve them with no delay. The second step in such an analysis is to see how the total delay over the rush hour depends upon the service rate μ.

For any given $A(t)$, the queue lengths, delays, etc., are quite sensitive to the service rate. One can see this immediately by observing how $D_q(t)$ would change if its slope were changed.

If the period of time over which a queue exists (t_0 to t_3 of Fig. 2.3(a)) is so large that $A(t)$ cannot be approximated by any simple formula, the easiest way to evaluate the total delay over the rush hour as a function of μ is simply to draw several curves of $D_q(t)$ for a reasonable selection of μ values. Then evaluate the area between $A(t)$ and each $D_q(t)$ using a planimeter or by counting squares on the graph paper.

If, however, the costs of delays are sufficiently large, one will eventually build a facility such that $D(t)$ is rather close to $A(t)$ with some small value of $t_3 - t_0$. Graphical methods are not very convenient in this case unless one can somehow make a preliminary guess of the likely range for the final choice of μ and therefore t_0 and t_3. To obtain any accuracy from graphical methods, one must draw the graph so as to magnify the range of t from t_0 to t_3. Instead of doing this, however, one might consider an analytic approach.

Suppose that $\lambda(t)$ rises to a maximum at a time t_1 as shown in Fig. 2.3(b). In most practical situations, it would be reasonable to assume that $\lambda(t)$ has a Taylor series expansion in $t - t_1$, or is at least twice differentiable near $t = t_1$, so that $\lambda(t)$ can be approximated by a quadratic function.

$$\lambda(t) = \lambda(t_1) - \beta(t - t_1)^2 \qquad (2.3)$$

for some constant β;

$$\beta = -\frac{1}{2} \frac{d^2\lambda(t)}{dt^2}\bigg|_{t=t_1},$$

at least over some sufficiently small range of $t - t_1$ (presumably for $t_0 < t < t_3$).

If μ is sufficiently close to $\lambda(t_1)$, the time t_0 of Fig. 2.3(b), where $\mu = \lambda(t_0)$, can be estimated from (2.3); also the time t_2 where μ is again equal to $\lambda(t)$.

$$\mu = \lambda(t_1) - \beta(t_0 - t_1)^2,$$

$$t_0 = t_1 - \left[\frac{\lambda(t_1) - \mu}{\beta}\right]^{1/2},$$

$$t_2 = t_1 + \left[\frac{\lambda(t_1) - \mu}{\beta}\right]^{1/2}. \tag{2.4}$$

It is convenient now to write $\lambda(t) - \mu$ in the factored form

$$\lambda(t) - \mu = \beta(t - t_0)(t_2 - t). \tag{2.4a}$$

This representation of $\lambda(t) - \mu$ follows directly from the postulates that it is quadratic in t, it has zeros at $t = t_0$ and $t = t_2$, and the second derivative with respect to t is -2β.

The queue at any time $t_0 < t < t_3$ is obtained by substitution of (2.4a) into (2.2),

$$Q(t) = \beta(t - t_0)^2 \left[\frac{(t_2 - t_0)}{2} - \frac{(t - t_0)}{3}\right]. \tag{2.5}$$

From this we see that $Q(t)$ grows quadratically in $t - t_0$ for t near t_0 and has a maximum at t_2,

$$Q(t_2) = \frac{\beta}{6}(t_2 - t_0)^3 = \frac{4[\lambda(t_1) - \mu]^{3/2}}{3\beta^{1/2}} \tag{2.6}$$

proportional to the 3/2 power of the oversaturation $\lambda(t_1) - \mu$. The queue vanishes again at time t_3 given by

$$t_3 = t_0 + (3/2)(t_2 - t_0) = t_0 + 3(t_1 - t_0). \tag{2.7}$$

Thus (2.5) can also be written as

$$Q(t) = \frac{\beta}{3}(t - t_0)^2(t_3 - t). \tag{2.5a}$$

Despite all the parameters, this function has a universal shape. If we were to translate the time and rescale the graph so that time was measured from t_0 in units of $t_3 - t_0$, and $Q(t)$ were measured in units of $Q(t_2)$, we could define

$$t' = (t - t_0)/(t_3 - t_0), \quad Q'(t) = Q(t)/Q(t_2),$$

and obtain

$$Q'(t) = (27/4)t'^2(1-t') \quad \text{for} \quad 0 \le t' \le 1.$$

This contains no parameters; it is a function only of t' which vanishes quadratically at $t' = 0$, linearly at $t' = 1$ and has a maximum of $Q'(t) = 1$ at $t' = 2/3$.

Finally, the total delay W over the rush hour is obtained from the area between $A(t)$ and $D_q(t)$, i.e., by integration of (2.5a).

$$W = \int_{t_0}^{t_3} d\tau Q(\tau) = \frac{\beta}{3} \int_{t_0}^{t_3} d\tau (\tau - t_0)^2 (t_3 - \tau)$$

$$= \frac{\beta}{3}(t_3 - t_0)^4 \int_0^1 du\, u^2 (1-u) = \frac{\beta}{36}(t_3 - t_0)^4$$

$$= \frac{9[\lambda(t_1) - \mu]^2}{4\beta}, \tag{2.8}$$

which is proportional to the square of the amount of oversaturation, $\lambda(t_1) - \mu$, or the fourth power of the duration of the queue $t_3 - t_0$. The β represents the curvature of $\lambda(t)$; a large β means a sharp peak for $\lambda(t)$, a small β a flat peak. A sharp peak, for fixed $\lambda(t_1)$ and μ, implies from (2.4) a short duration of the queue, a small maximum queue (2.6), and a small total delay (2.8).

The estimation of total delay is one of the most common problems to arise in practical applications. Some further refinements of it will be discussed again in Chapter 8. Any conclusions obtained here are tentative and subject to unknown errors arising from the use of deterministic approximations. One must be particularly cautious of the possibility that the queue lengths calculated here may be overshadowed by queues generated by stochastic effects.

2.4   Delays over many years

For many types of service facilities, the arrival rate of customers shows rush hours each day, variations in 24-hour arrival patterns throughout the week, seasonal variations, and a general growth in demand from year to year as discussed in Section 2.1. A question that often arises is the following: one has an estimate of the annual growth of demand over the next several years and one knows the costs of construction of facilities of various service rates μ; when should one expand the service?

We are concerned here mainly with the method of evaluating the delays associated with various strategies rather than with the final problem of selecting an optimal strategy. In principle, one can draw a graph of $A(t)$ for a time range of 5 years, but if one draws it on a 5-year time scale, one does not see the individual daily rush hours which are the source of the delays; a 5-year plot will show the annual growth, possibly the seasonal oscillation, but little else.

Since the queue is likely to vanish at the same time each day, one could draw separate graphs of the $A^{(j)}(t)$ for the jth day, and represent the total delay as the sum over j of the areas between the $A^{(j)}(t)$ and $D_q^{(j)}(t)$. Although one could, in principle, draw some 10^3 such graphs, and evaluate them separately, it should be possible to classify the days into weekends, weekdays, etc. Days of the same classification are still likely to show trends or growth, but the shapes of the $A^{(j)}(t)$ are likely to be nearly the same, differing only in the total count (scale). Formally, for days of the same class, we might assume that

$$A^{(j)}(t) = A^{(j)}F(t) \qquad (2.9)$$

in which $A^{(j)}$ is independent of t and $F(t)$ is independent of j. One could, for example, let $A^{(j)}$ be the 24-hour count so that $F(24) = 1$.

If we draw a graph of the function $F(t)$ as in Fig. 2.4(a), we can

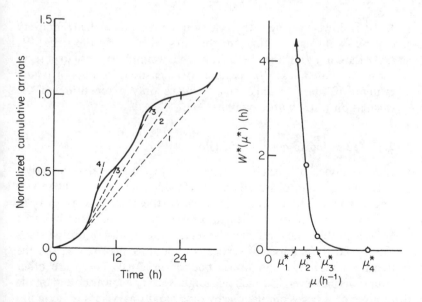

Figure 2.4 *Evaluation of a rescaled wait for various service rates*

evaluate the total wait (area) $W^*(\mu^*)$ that would exist for a hypothetical cumulative arrival curve $F(t)$ and service rate μ^*. If we do this for many value of μ^*, we can construct a graph of $W^*(\mu^*)$ versus μ^* as in Fig. 2.4(b).

The delay that would occur for an arrival curve $A^{(j)}F(t)$ and service rate μ can now be obtained simply by rescaling coordinates in Fig. 2.4. The delay $W^{(j)}$ is just $A^{(j)}$ times the W^* evaluated at a service rate $\mu^* = \mu/A^{(j)}$, i.e.,

$$W^{(j)} = A^{(j)}W^*(\mu/A^{(j)}). \qquad (2.10)$$

Actually, it may be more advantageous to draw a graph of the function

$$H(s^*) = s^*W^*(1/s^*) \qquad (2.11)$$

rather than $W^*(\mu^*)$, because (2.10) can then be written as

$$W^{(j)} = \mu H(A^{(j)}/\mu). \qquad (2.12)$$

The total delay over many days (years) having the same $F(t)$ is obtained by adding the $W^{(j)}$ for each day,

$$\text{total delay} = \mu \sum_j H(A^{(j)}/\mu). \qquad (2.13)$$

If $A^{(j)}$ is slowly varying with j, one would not evaluate the H for every day but would group together the days with nearly the same $A^{(j)}$. Once the curve $H(s^*)$ has been drawn, it is a simple exercise to observe the H evaluated at $A^{(j)}/\mu$, i.e., the $A^{(j)}$ measured in units of μ, and evaluate (2.13). A change in μ involves only a repetition of the calculation (2.13) with new units.

2.5 Queueing to meet a schedule

In most models of queueing it is customary to imagine that the arrival pattern is given or observed and that it will not change if the service rate changes. The objective is to serve these arrivals as early as possible, i.e., to maximize the number of service completions by time t subject to the constraint that the service rate shall not exceed some rate μ and the number of service completions shall not exceed the number of arrivals (an upper bound on $D(t)$). There are other situations, however, in which one might specify a lower bound on the number of service completions by time t and one wishes to minimize the number of service completions by time t subject to the same constraint that the service rate not exceed μ and that the number of service completions stays above the lower bound.

Suppose, for example, that a factory can manufacture goods at some maximum rate μ. The future demand for these goods is predictable; $D_d(t)$ goods should be produced by time t (the cumulative demand). The demand rate $dD_d(t)/dt$ is time dependent (rush hours, seasonal demand, etc.) and may at time exceed μ, although the long time average demand rate is less than μ (i.e., for sufficiently large t, $D_d(t) < \mu t$).

To meet this demand it is necessary to stockpile goods ahead of the demand surges, but one does not wish to store any more than necessary. The number of goods produced by time t, $D(t)$, should therefore be the minimum number such that $D(\tau) \geq D_d(\tau)$ and $dD(\tau)/d\tau \leq \mu$ for all values of τ including $\tau > t$.

The method of constructing $D(t)$ is similar to the construction of $A(t)$ in the conventional queueing problem except that one starts at some time in the distant future and draws a line of slope μ backwards in time whenever $dD_d(t)/dt$ exceeds μ as illustrated in Fig. 2.5. It is in fact the same type of construction as for $A(t)$ if, for some future time t^* when the demand will certainly be satisfied, one draws a graph of $D_d(t^*) - D(t^* - \tau)$ versus $t^* - \tau$, the future work to be done (by time t^*) versus the time remaining to do it, $t^* - \tau$.

Of course, vertical and horizontal distances between $D(t)$ and $D_d(t)$ have the obvious interpretations as the stockpile of goods and the time any object remains in inventory (if goods are used in the order they are produced).

Another example of the same type of theory relates to the morning commuter rush hour. Suppose that a transportation system has a bottleneck which can accommodate a maximum flow of μ. Let $D_d(t)$ represent the number of commuters (customers) who must be at work by time t, or actually the number that must have passed the bottleneck by time t in order to be at work on time, and suppose that $dD_d(t)/dt > \mu$ during some time interval. In order for everyone to be at work on time it is necessary that some people arrive at work ahead of schedule.

If some transportation manager could assign arrival time reservations, he would presumably tell the persons who should be at work at time t to arrive at time τ such that $D(\tau) = D_d(t)$ as in Fig. 2.5 much as a factory manager would produce $D(\tau)$ goods by time τ in order to satisfy the demand $D_d(t)$ at time t.

Unfortunately, most transportation facilities do not have a reservation system and a traveler whose aim is to maximize his own benefit

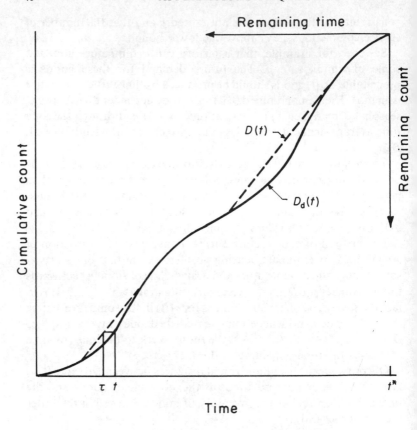

Figure 2.5 *Cumulative counts to meet a scheduled demand*

is not likely to cooperate. If there were no queue behind the bottleneck, a person who must be at work by time t would see that he could be at work on time even if he arrived at late as time t, but someone else would be late for work. This is one of many examples in transportation and elsewhere in which minimizing some global objective is not achieved by each person doing what is optimal for himself.

One might now ask if there is some arrival curve $A(t)$ which results if every persons tries to do what is optimal for himself. An arrival pattern which identifies an arrival time with each individual would be described as *stable* if, for that pattern, no individual can find a better arrival time than the one he has. An interesting feature of this situation is that there is *no* arrival pattern with finite arrival rate and FIFO queue discipline that is stable with respect to the objective that each individual arrives as late as possible so as still to be at work on time.

To prove this, one need only observe that any assignment giving an arrival curve A with finite slope as in Fig. 2.6 and guaranteeing that each person is at work on time will necessarily cause some people to be at work early. However, any such person who must be at work by time t can determine from the curve A an arrival time t' as shown in Fig. 2.6 such that he will be at work exactly at time t, an arrival time which, for some people at least, must be later than the one assigned (thus the assignment is not stable).

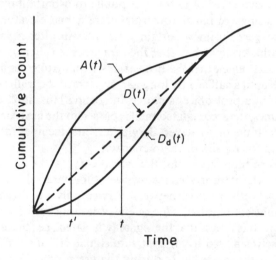

Figure 2.6   *Arrivals who must be at work on time*

To allow simultaneous arrivals (infinite slope for A) does not resolve the difficulty unless there is some mechanism by which people who arrive simultaneously can be served in the order of their work schedules. Neither is it helpful to recognize that there may be some restriction upstream of the bottleneck which limits the rate of arrivals, because the curve A actually refers to the expected times of arrivals, the times people would arrive in the absence of a queue. It makes no difference where they wait.

To obtain a stable assignment, one must postulate a different type of individual objective than arriving as late as possible. Stable

assignments do exist under the hypothesis that each individual assigns
a price p per unit of delay in queue and a price $p' < p$ per unit of time
by which he arrives to work early. He then selects an arrival time so as
to minimize his cost. An assignment is stable if no individual can find a
less costly arrival time than the one he has.

2.6 Pulsed service

For many types of service facilities, service occurs in pulses. A traffic
signal at a highway intersection passes cars at a fairly constant rate for
a certain time while the signal is green, but then provides no 'service'
while the signal is red. Any form of public transportation providing
service at a single terminal (an airplane, train, bus, elevator, etc.) will
load passengers during a certain time interval after a vehicle has
arrived at the terminal; but after the vehicle leaves, there is no service
until the next vehicle arrives. Pedestrians wishing to cross a highway
will queue until a sufficiently long gap appears in the traffic stream of
cars. Mail in a post office is stored in sorting bins until, at certain
discrete times, the accumulated mail is passed to the next sorter or put
on a truck, train, or whatever. Service at some facilities may also be
interrupted for repairs or maintenance.

In many of these situations, the arrival curve $A(t)$ is smooth in the
sense that $\lambda(t)$ is nearly constant over time intervals of duration
comparable with the time between service pulses and each service
pulse is sufficient to exhaust the queue of waiting customers. For the
traffic signal this means that the signal is 'isolated', i.e., the arrivals are
not themselves pulsed by an upstream signal, and the traffic is light
enough that the queue clears during the green time. For the public
transportation example, the vehicles have sufficient capacity to serve
all waiting customers. The pedestrians cross a street in a pack
whenever there is a gap, and mail pick-ups take all the mail which has
accumulated.

Deterministic fluid approximations are particularly useful for a
crude analysis of these types of situations. The queues may well
become sufficiently large between service pulses to justify use of such
approximations even though, over a long time period, the system is
'undersaturated' in the sense that the queue does not continue to grow
from one cycle to the next.

A typical graph of arrivals and departures for the above type of
system is shown in Fig. 2.7. Suppose, as suggested by the traffic signal

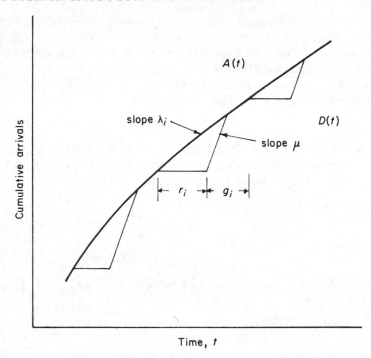

Figure 2.7 *Arrivals and departures for pulsed service*

application, we let

$$r_i = \text{red time of the } i\text{th cycle}$$
$$g_i = \text{green time of the } i\text{th cycle}$$
$$\lambda_i = \text{arrival rate during the } i\text{th cycle}$$
$$\mu = \text{service rate during green.}$$

Since $\lambda_i < \mu$, the maximum wait in the ith cycle is r_i. If N_i is the number of customers delayed in the ith cycle, then the queue clears at time

$$N_i/\lambda_i = N_i/\mu + r_i,$$

so

$$N_i = r_i \lambda_i / (1 - \lambda_i/\mu).$$

The total wait in the ith cycle, the area of the ith triangle in Fig. 2.7, is therefore $r_i N_i/2$, i.e.,

$$\text{wait in } i\text{th cycle} = \frac{1}{2} \frac{r_i^2 \lambda_i}{(1 - \lambda_i/\mu)}. \qquad (2.14)$$

We have assumed here that $\lambda(t)$ is nearly constant over the ith cycle with a value λ_i. It should also be nearly the same for adjacent cycles, but we do not rule out the possibility that λ_i may change appreciably over a sufficiently long time. We are making no restrictions, however, on how the g_i and r_i vary from cycle to cycle, as long as g_i is large enough for the queue to clear in each cycle, i.e.,

$$\lambda_i(r_i + g_i) \le \mu g_i. \tag{2.15}$$

According to the definitions of Section 1.4, the average wait per customer during the ith cycle is the total wait during the ith cycle divided by the number of customers in the ith cycle:

$$\text{average wait per customer in the } i\text{th cycle} = \left[\frac{r_i^2 \lambda_i}{2(1 - \lambda_i/\mu)}\right]\left[\frac{1}{\lambda_i(r_i + g_i)}\right]$$

$$= \frac{r_i^2}{2(r_i + g_i)(1 - \lambda_i/\mu)}. \tag{2.16}$$

Correspondingly, the average wait per unit time, i.e., the average queue length is

$$\text{average queue during the } i\text{th cycle} = \frac{r_i^2 \lambda_i}{2(r_i + g_i)(1 - \lambda_i/\mu)}. \tag{2.17}$$

Over n cycles, the average wait is again the total wait during n cycles divided by the number of arrivals over n cycles.

average wait per customer over cycles $j + 1$ to $j + n$

$$= \langle w \rangle = \frac{\frac{1}{2}\sum_{i=j+1}^{j+n} r_i^2 \lambda_i/(1 - \lambda_i/\mu)}{\sum_{i=j+1}^{j+n}(r_i + g_i)\lambda_i}$$

$$= \frac{\frac{1}{n}\sum_{i=j+1}^{j+n} \frac{1}{2}r_i^2 \lambda_i/(1 - \lambda_i/\mu)}{\frac{1}{n}\sum_{i=j+1}^{j=n}(r_i + g_i)\lambda_i}. \tag{2.18}$$

We have divided numerator and denominator by n, because the numerator has the interpretation as the arithmetic average wait per cycle and the denominator the arithmetic average number of arrivals per cycle time.

The important thing to observe here is that (2.18) is not generally the same as the arithmetic average of (2.16).

If λ_i varies sufficiently slowly with i that it is nearly constant over n cycles, (2.18) simplifies to

$$\langle w \rangle \simeq \frac{(1/n) \sum_{i=j+1}^{j+n} r_i^2}{2(1 - \lambda_j/\mu)(1/n) \sum_{i=j+1}^{j+n} (r_i + g_i)}. \tag{2.19}$$

In many of the above applications, the service in each cycle ceases as soon as the queue vanishes. This is approximately true for a vehicle-actuated signal and for passengers boarding a public transportation vehicle. In this case, the equality holds in (2.15) for all i. One can exploit this to eliminate one of the variables r_i or g_i or to express both in terms of the total cycle time $(r_i + g_i)$. If we do the latter (2.14) becomes

$$\text{wait in } i\text{th cycle} = \frac{1}{2}(1 - \lambda_i/\mu)(r_i + g_i)^2 \lambda_i$$

and (2.19) becomes

$$\langle w \rangle \simeq \frac{1}{2}(1 - \lambda_i/\mu) \frac{(1/n) \sum_{i=j+1}^{j+n} (r_i + g_i)^2}{(1/n) \sum_{i=j+1}^{j+n} (r_i + g_i)}. \tag{2.20}$$

In some applications, the cycle time varies appreciably from cycle to cycle (even though λ_i does not). A vehicle-actuated signal, for example, may have a red time determined by fluctuating traffic from a side street, so that $(r_i + g_i)$ is interpreted as a random variable with some specified probability distribution. For a public transportation system, the time $(r_i + g_i)$ is interpreted as the 'headway' between dispatches which may also have random fluctuations. For pedestrians crossing a street, $(r_i + g_i)$ is the time interval between acceptable gaps in the traffic stream.

If we interpret $(r_i + g_i)$ as an ith observation of a random variable $(R + G)$, then the arithmetic mean of many observations is (by definition) interpreted as the expectation. Thus

$$(1/n) \sum_i (r_i + g_i)^2 \simeq E\{(R+G)^2\} = E^2\{R+G\} + \text{Var}\{R+G\},$$

$$1/n \sum_i (r_i + g_i) \simeq E\{R+G\}$$

and

$$\frac{(1/n) \sum_i (r_i + g_i)^2}{(1/n) \sum_i (r_i + g_i)} \simeq \frac{E^2 \{R + G\} + \mathrm{Var}\{R + G\}}{E\{R + G\}}$$

$$= E\{R + G\} [1 + C^2 (R + G)]$$

in which C is the coefficient of variation

$$C^2 (R + G) \equiv \mathrm{Var}\{R + G\}/E^2 \{R + G\}. \tag{2.21}$$

In many cases (particularly for loading bus passengers), the service rate is large compared with λ_i, i.e. $\lambda_i/\mu \ll 1$ and $g_i/r_i \ll 1$. Thus (2.21) simplifies still further to

$$\langle w \rangle \simeq \frac{1}{2} E\{R\} [1 + C^2 (R)]. \tag{2.22}$$

If headways between buses, for example, are all equal (so that $C^2 (R)$ = 0), it is obvious that a person who arrives at a random time unrelated to any possible bus schedule will wait, on the average, a half a headway, i. e., $\frac{1}{2} E\{R\}$. If, however, the headways are irregular, it is more likely that a person will arrive during a long headway than during a short headway; therefore, the average wait will be larger than that associated with the average headway. In the absence of any control to keep buses on schedule, there is, typically, a tendency for the headway distribution to become exponential. For an exponential distribution $C^2 = 1$, i.e., the average wait for buses with an exponential headway distribution is twice that of a regular schedule with the same average headway.

2.7 Applications

In most applications of the above formulas, the choice of the time between pulses is subject to certain constraints. Clearly, if one could freely choose any cycle time or headway, one would choose it to be arbitrarily small so as to minimize delay.

(a) *A fixed-cycle traffic signal*
An isolated traffic signal actually serves two (or more) traffic streams, each of which receives pulsed service. During part of the red time for one stream, the signal is green for the other traffic stream. There is, however, an effective lost time when neither traffic stream flows.

If r_i' and g_i' represent the red and green times for the second traffic stream, we can write

$$r_i = g_i' + L \quad \text{and} \quad r_i' = g_i + L,$$

in which L is the effective lost time per cycle. The cycle time is given by

$$r_i + g_i = r_i' + g_i' = g_i + g_i' + L.$$

For just two traffic streams with arrival rates λ_i and λ_i' and service rates μ and μ', the wait per cycle is, according to (2.14)

$$\text{wait in } i\text{th cycle} = \frac{(g_i' + L)^2 \lambda_i}{2(1 - \lambda_i/\mu_i)} + \frac{(g_i + L)^2 \lambda_i'}{2(1 - \lambda_i'/\mu')}$$

provided

$$\lambda_i (g_i + g_i' + L) < \mu g_i \quad \text{and} \quad \lambda_i'(g + g_i' + L) < \mu' g_i'. \tag{2.23}$$

If the arrival rates are constant, $\lambda_i = \lambda$, $\lambda_i' = \lambda'$, and the signal is periodic, $g_i = g$, $r_i = r$, etc., then the wait per unit time is

$$\frac{(g + L)^2 \lambda}{2(g + g' + L)(1 - \lambda/\mu)} + \frac{(g' + L)^2 \lambda'}{2(g + g' + L)(1 - \lambda'/\mu')} \tag{2.24}$$

provided (2.23) is true; otherwise the queues grow from cycle to cycle. In (2.23), λ/μ can be interpreted as the fraction of the cycle time necessary to serve the flow λ. A necessary condition for (2.23) to hold is that

$$\frac{\lambda}{\mu} + \frac{\lambda'}{\mu'} < 1 - \frac{L}{(g + g' + L)}. \tag{2.25}$$

Typically a traffic engineer has the option of choosing the g and g' within some practical range of values satisfying (2.23). He might choose them so as to minimize (2.24). For most reasonable values of the parameters, the minimum of (2.24) occurs at the boundary (2.23), i.e., for the shortest possible g and g' which can accommodate the flows. The issues of traffic signal setting are more complex than this, however, because stochastic effects become very important as the g and g' approach their minimum values. If one includes the delays due to stochastic effects, the minimum delay per unit time actually occurs for a cycle time approximately twice that predicted above. The deterministic theory, however, describes at least a first (but very crude) approximation to the delays. Some of the stochastic effects will be discussed in Chapter 9.

In practice, the values of g and g' are usually constrained also by the time it takes a pedestrian to cross the road.

(b) *Bus dispatching*
In selecting the headways between buses, it is clearly advantageous to make $E\{R\}$ and $C^2(R)$ as small as possible. Since buses are subject to various random disturbances generated by traffic congestion, signals, loading times, etc., it is usually necessary to introduce some types of control in order to keep $C^2(R)$ small. The most common scheme of regulation is to impose a schedule.

It is easy to prevent buses from running ahead of schedule but more difficult to prevent them from running late. To obtain a two-sided control, one must provide some slack time in the schedule so that buses can gain on the schedule once they have fallen behind. The more slack one introduces, however, the slower the speed and, for a fixed number of buses, the longer is the average headway. The minimum average wait per passenger involves a compromise between a small $E\{R\}$ (loose control) and a small $C^2(R)$ (tight control). The best strategy will not generally involve exactly equal headways, $C^2(R) = 0$.

If one controls headways so that $C^2(R) \ll 1$, the issue of selecting an optimal $E\{R\}$ usually centers around the fact that there is a cost (per unit time or per trip) associated with each bus dispatched. In principle, this cost should be balanced against the cost or inconvenience of delay. If it costs γ to despatch a vehicle and it costs p per unit of wait, the cost per unit time of bus operation is $\gamma/E\{R\}$ and the total cost per unit time (for $C^2(R) \ll 1$) is approximately

$$\frac{\gamma}{E\{R\}} + p\frac{E\{R\}}{2}\lambda(t). \tag{2.26}$$

If $\lambda(t)$ varies only slightly during a headway, the minimum total cost over a long period of time can be achieved by choosing $E\{R\}$ so as to minimize the cost rate (2.26) at every t. Thus the optimal $E\{R\}$ at time t is

$$E\{R\} = \left[\frac{2\gamma}{p\lambda(t)}\right]^{1/2}. \tag{2.27}$$

With this choice of $E\{R\}$, the costs of delay and of operation are equal.

The number of passengers on each bus is

$$\lambda(t)E\{R\} = \left[\frac{2\gamma}{p}\lambda(t)\right]^{1/2}. \tag{2.28}$$

If over some long period of time $\lambda(t)$ should increase by a factor of 4 (but γ and p remain constant), the increased passengers would be accommodated by dispatching twice as many buses with each bus carrying twice as many passengers.

If buses have limited capacity and (2.28) should exceed the capacity of the bus, the optimal strategy is to dispatch the buses so that they are barely full. If one were to dispatch at any longer headway, passengers would be left behind and the delays would grow arbitrarily large. It is generally true that once a vehicle is full, it should be dispatched immediately. Nothing can be gained by having it sit waiting for passengers who cannot board.

(c) Queueing for gaps

For pedestrians crossing a highway or cars which must yield to another traffic stream, the delays to these customers depend upon the time interval between acceptable gaps. The time between services is, of course, quite sensitive to the arrival rates of the opposing traffic stream and its headway distribution.

The typical question which arises here is whether or not one should install a traffic signal which, in effect, changes the headway distribution on the opposing stream. As in example (a), the issue now becomes a balance between the delays for the two traffic streams.

Problems

2.1 A service facility (highway) is capable of serving customers (cars) at a constant rate of μ customers per hour. Customers arrive at a constant rate $\lambda_1 < \mu$ until some time $t = 0$ (7.00 a.m.), but from $t = 0$ until some time τ (9.00 a.m.) they arrive at a rate $\lambda_2 > \lambda_1$. After time τ the arrival rate returns to the value λ_1 and remains there until time τ' (7.00 a.m. the next day) when the pattern repeats itself. If $\lambda_2 > \mu$, customers who cannot be served immediately form a queue and are served first-in, first-out.

Draw curves for the cumulative arrivals and departures of customers starting at time $t = 0$ when there is no queue. Evaluate and identify geometrically

(a) the maximum queue length
(b) the longest delay to any customer
(c) the duration of the queue
(d) the total delay to all customers during the time 0 to τ'.

2.2 As in Section 2.2, let $A(t)$ represent the cumulative number of customers to arrive by time t and $D_q(t)$ the cumulative number to enter the service. Suppose, however, that there is a storage space for only c customers (enough to keep the server busy at all times when $\lambda(t) > \mu$); anyone who arrives when the storage is full goes away (he may be served elsewhere). If the server can serve at a maximum rate μ and $A(t)$, $\lambda(t)$ are as shown in Fig. 2.3(a), (b), determine $D_q(t)$ and $\mu(t)$.

2.3 In selecting a facility to serve the arrival pattern of problem 2.1, a designer has the option of choosing any values of μ. The cost per unit time of providing a service rate μ (labor, interest on investment, etc.), however, is proportional to μ, i.e.,

$$\text{service cost during time } \tau' = \alpha\mu\tau', \ \alpha = \text{constant}$$

regardless of whether or not the facility is fully utilized. The designer proposes that the value of a customer's time is worth p per unit time ($\$$ per hour), i.e., the total cost of delay is p times the total delay.

Determine the choice of μ which the designer would select if his objective is to minimize the sum of service cost and delay costs.

2.4 Suppose that on day j the arrival curve of customers to a facility with fixed service rate μ has the form

$$A_j(t) = A_j F(t)$$

with $F(t)$ independent of j. On day 0 the arrival rate has a single maximum of the type

$$\lambda_0(t) = \mu - \beta(t - t_1)^2$$

i.e., the maximum arrival rate on day 0 is just equal to the service rate.

If the demand increases at a constant fractional rate of α per day

$$A_j = A_0[1 + \alpha j] \quad \text{for} \quad j \geq 0$$

how will the total delay W_j on day j increase with j.

2.5 An automobile assembly plant can assemble cars only at a single rate of μ cars per day or close down. It costs $\$p$ per day to store an assembled car. There is a fixed cost per day that is independent of whether the factory is operating or not. In addition, there is a cost of $\$\alpha$ per day (labour) to operate at rate μ and the equivalent of 2 weeks of operating cost to close the factory and start again (no matter how long it is closed).

What strategy of operation should be used to minimize the long time average cost per day of operation if the factory must satisfy a steady demand of λ cars per day, $\lambda < \mu$?

2.6 A vehicle-actuated traffic signal serves two traffic streams with cumulative arrival curves $A_1(t)$ and $A_2(t)$. Show how one could graphically construct departures curves $D_1(t)$ and $D_2(t)$ for the two traffic streams if the queue discipline alternates as follows: stream 1 is served at a rate μ until the queue in stream 1 vanishes, then there is a lost time L during which no one is served, then stream 2 is served at a rate μ until that queue vanishes, then is another lost time L followed by service to stream 1, etc.

2.7 Two bus routes, one with headways of 10 minutes, another with headways of 20 minutes, merge along a common section of route. The schedules are synchronized so as to create headways in the sequence 5, 5, 10, 5, 5, 10, ... If a passenger who wishes to travel along the common route arrives at a random time, what is the probability that he must wait for a time greater than t? What is his average waiting time?

2.8 Each of two buses carries passengers from a depot to various destinations and return for another trip with a round trip time very nearly equal to T. The buses are run by independent drivers, however, who make no attempt to coordinate their schedules. Actually, one bus runs slightly faster than the other so that over many trips the fraction of trips that the second bus leaves within a time t after the first bus is t/T, $0 < t < T$. In effect, the times between departures of the buses are random with a uniform distribution over the interval $0 < t < T$.

If passengers arrive at the depot at a constant rate, what is the average time that a passenger must wait for the next bus? Compare this with the wait if the headways are controlled so as to be $T/2$.

2.9 Two shuttle buses, each of which can carry c passengers, serve the same bus depot. The time between dispatches of the same bus is 15 minutes, but one of the buses is dispatched 10 minutes after the other so as to create headways 10, 5, 10, 5, etc. Passengers arrive at the depot at a constant rate of λ per minute.

If any passenger who arrives at the depot and finds a queue of c passengers goes away, what is the long time average number of passengers served per unit time by the buses, as a function of λ? What is the average wait per passenger for those passengers who actually board a bus?

2.10 Passengers arrive at a bus stop A at a constant λ, to be transported to point B. A bus company has two buses assigned to route A–B; one of them has a capacity of 30 passengers, the other a capacity of 50 passengers. Both vehicles have the same speed and can make the round trip in 10 minutes plus an extra 6 seconds total for loading and unloading of each passenger carried. The vehicle unloads at B and returns to A empty.

Determine as a function of λ, the strategy for dispatching vehicles from A which will minimize the average wait per passenger; λ is less than the sum of the maximum service rates of the two buses. Disregard any transient effects due to the starting conditions or effects due to the integer valued count of passengers.

Note: there are three ranges of λ. As λ increases, the first two ranges of λ have well defined solutions. For the largest values, the exact solution is very complicated.

CHAPTER 3

Simple queueing systems

3.1 Introduction

Most service systems involve many servers, many queues, and many different types of customers, or objects which satisfy a conservation principle. These are likely to be interrelated in the sense that the operation of one component depends upon the others.

In the analysis of a complex queueing system (cars moving along a freeway system, passengers moving through an airport terminal, etc.) it is advantageous first to identify *all* objects (customers, servers, money, etc.) which satisfy conservation principles. Next construct a 'schematic flow chart' showing how objects of each type move from one service point to another. Finally, at each ith service point and for each jth object type, one may identify the existence of an arrival curve A_{ij}, a departure curve D_{ij} (or possibly both a D_{qij} and a D_{sij}, or a D^*_{qij} and a D^*_{sij} if the order of service is relevant) and corresponding queue lengths, delays, etc., for the jth object type at the ith service point. The actual construction of these curves will depend upon the description of the service mechanism, interactions between objects, etc., but, usually, between a flow chart and a collection of cumulative flow graphs one will have, in principle, a complete description of everything one would care to know about the evolution of the system.

It is not possible to describe a universal procedure designed to solve all problems. Elegant methods of solution of particular problems usually exploit special features of the particular problem at hand and lay special emphasis on those aspects of the system behavior which are of most practical importance. We will consider here a few examples of hypothetical situations, so as to give some indication of how one might analyze fairly complex systems by a sequence of elementary graphical manipulations.

3.2 Series or tandem queues

In many service systems a customer passes through a succession of service facilities having specified rules of behavior. A car, for example,

joins a queue at one traffic signal, moves on to join another queue at
the next intersection, and so on. At an airport terminal, a passenger
first must check baggage at one service counter, join a queue to enter
the departure lounge, wait in the departure lounge, and then join a
queue to pass through the door of the aircraft. On an assembly line, a
production item passes from one service to another until it finally
emerges as a finished product. It may, at any stage, join a storage, i.e., a
queue.

In the simplest systems of this type, we have only one customer type
but several service points. The schematic flow chart is as shown in Fig.
3.1(a). Customers enter the system, join a queue at storage 1, pass

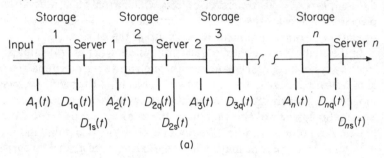

(a)

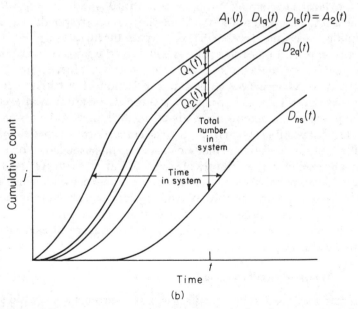

(b)

Figure 3.1 *Graphical representation of tandem queues*

through server 1, join a queue at storage 2, pass through server 2, etc., finally pass through server n and leave the system. The special structural feature of this system is that there are no exits from or entrances into the system except at 1 and n. Any customer which passes the first server must eventually pass all other servers.

To describe the evolution of the system, we might imagine (as in Section 1.2) that we place observers at the entrance to each storage, the exit from each storage, and possibly also the exit from each server. If one is concerned with the possibility that customers may pass each other somewhere in the system, we·could ask the first observer to assign consecutive numbers to the customers which pass him. If they pass him at times

$$0 \le t_{11} \le t_{12} \le t_{13} \le \dots$$

he can construct a curve

$$A_1(t) = \text{cumulative number of customers to enter the system.}$$

We assume that the system is empty at time 0 or, if not, that observer 1 assigns numbers to all customers in the system at time 0 and arbitrarily records them as having passed him at time 0.

As described in Section 1.2, an observer at the exit to the first queue can record the times

$$t^*_{1qj} = \text{time customer } j \text{ leaves the first storage}$$

or the times

$$t_{1qj} = \text{time of the } j\text{th ordered departure from the first storage,}$$

and represent these data graphically as in Fig. 1.4 or 1.3 by curves D^*_{1q} and D_{1q}, respectively. An observer at the exit to the first server can similarly record the times t^*_{1sj} at which customer j passes him or the corresponding ordered times t_{1sj}, and draw curves D^*_{1s} and D_{1s}. Observers at the second, third, etc., servers can also record times t^*_{2j} or t_{2j}, t^*_{2qj} or t_{2qj}, t^*_{2sj} or t_{2sj}, etc., at which customers pass them, and draw curves A^*_{2q} or A_2, D^*_{2q} or D_{2q}, etc.

How the above data are recorded does not depend upon what lies between the various observers. Indeed, what the last observer sees is independent of whether or not there were other observers at intermediate points. Since

$$t_{1j} \le t^*_{1qj} \le t^*_{1sj} \le t^*_{2j} \le \dots$$

and

$$t_{1j} \leq t_{1qj} \leq t_{1sj} \leq t_{2j} \leq \ldots$$

none of the curves $A_1, D_{1q}^*, \ldots$ or $A_1, D_{1q}, \ldots$ intersect. One could superimpose them all on the same graph and thereby record all observations on the same figure. If customers pass each other, the curves D_{kq}^* or D_{ks}^* could be highly irregular. If, however, customers retain the same ordering, or we consider the curves D_{kq}, D_{ks}, the successive curves will be monotone and probably fairly smooth on a scale large compared with 1, as illustrated schematically in Fig. 3.1 (b).

On a diagram such as Fig. 3.1 (b) one can immediately identify most of the important characteristics of the system behavior. The vertical distance between any two curves at time t represents the number of customers between the appropriate observation points (and the sum of those between intermediate points). In particular, the vertical distance from D_{ns} to A_1 at time t is the total number of customers in the system at time t. Correspondingly, the horizontal distance between any D^* or A^* curves (equivalently D or A if customers stay in order) at height j represents the transit time of customer j between the two observers; in particular the horizontal distance from A_1 to D_{ns}^* is the time the jth customer spends in the system. The area between A_1 and D_{ns} or D_{ns}^* represents the total time spent in the system by all customers.

The above merely describes a method of recording observations. Any description of the various servers will presumably give a recipe from which one can sequentially evaluate D_{kq}^* and D_{ks}^* from A_k^* or D_{kq} and D_{ks} from A_k, just as if there were but a single server k serving an input A_k. If, for example, the kth server is a traffic signal on a highway, it may serve customers at a rate μ during specified green times, provided there is a queue, i.e., $D_{kq}(t) < A_k(t)$.

A description of the system would also specify a relation between A_{k+1}^* and D_{ks}^*. If, for example, a customer leaving server k immediately joins the queue for server $k+1$, then A_{k+1}^* and D_{ks}^* are the same curves. If each jth customer has a specified nonzero transit time from the kth server to the $(k+1)$th queue, this will, of course, determine A_{k+1}^* from D_{ks}^*. If, in particular, all customers have the same transit time τ, A_{k+1}^* is a horizontal translation of D_{ks}^* by the time τ.

3.3 Sorting of mail

As an illustration of tandem queues, we consider the flow of mail through a sequence of sorters in a post office. Here the 'customers' are letters and the servers are the mail sorters.

Oliver and Samuel (1962) analyzed (by somewhat different methods) the following problem. If there is a fixed total number of mail sorters, all of whom can sort mail at the same continuous rate at any of the sorting stages and the cumulative input of mail to the first of n sorting stages is given, how should the sorters be distributed among the n stages so as to (a) minimize the total letter delay through the n stages, or (b) maximize the cumulative output from the last stage at some specified dispatch time?

Apparently some post offices had operated under the following (incorrect) strategy. When a loading of mail comes in, put most of the sorters at the first stage until the queue vanishes, then shift the sorters to the second stage, etc. The sorters follow the surge of mail through the sorting stages.

The evolution of the system can be represented graphically as in Fig. 3.1 for any specified allocation of sorters, but we will neglect the transit time from one sorter to the next and also the number of letters actually in service. Let

$$D_k(t) \equiv D_{kq}(t) \simeq D_{ks}(t) \simeq A_{k+1}(t).$$

We assume that there are so many servers that we can disregard their discrete nature, and that the total service rate of μ for all servers combined can be partitioned in any way among the n stages. The allocation can also vary with time in any way without penalty. If $\mu_k(t)$ is the service rate of the kth stage, then

$$\sum_{j=1}^{n} \mu_k(t) = \mu.$$

The objective of minimum total delay through the system or maximum number of departures from the system depends explicitly upon the final departure curve from the last stage. It depends upon the departure curves through intermediate stages only in that the final departure is constrained by those ahead. The constraints are

$$\mu_k(t) = \frac{dD_k(t)}{dt} \geq 0, \quad \sum_{k=1}^{n} \frac{dD_k(t)}{dt} \leq \mu$$

and $D_n(t) \leq D_{n-1}(t) \leq \ldots \leq D_1(t) \leq A(t)$.

The objective, generally speaking, is to make $D_n(t)$ as large as possible. This suggests that we should make

$$D_n(t) = D_{n-1}(t) = \ldots = D_1(t) \leq A(t)$$

and operate all stages at a capacity μ/n. The whole system would then operate like a single-server queue with service rate μ/n.

There will be many strategies which give a maximum $D_n(t)$ for any single specified value of t. The above strategy, however, yields the maximum $D_n(t)$ simultaneously for all values of t. It, clearly, is never advantageous to operate in such a way as to form a queue between sorting stages because the excess service that was used to create the queue could have been used instead to serve that queue. Thus, each $\mu_k(t)$ should match the preceding one.

3.4 A continuum of service points in series

In the previous sections we clearly had in mind a system consisting of a succession of discrete service points, at each of which one had some rules for constructing $D_{kq}(t)$ from $A_k(t)$. Suppose, however, we place a succession of traffic counters at locations $x_1, x_2, \ldots$ along a stretch of highway having no side exits or entrances. Each counter starts counting from zero with the passage of some identifiable vehicle (a truck). Suppose also that this vehicle neither passes or is passed by any other vehicle over this section of highway.

Let $A_k(t)$ be the cumulative count to time t of cars passing the counter x_k. At any time t, the difference $A_k(t) - A_{k+1}(t)$ represents the number of vehicles which have passed x_k but have not yet passed x_{k+1}; it is the number of vehicles between x_k and x_{k+1} (a 'queue' between x_k and x_{k+1}). If no vehicles pass each other (FIFO service everywhere), the horizontal distance $A_{k+1}^{-1}(y) - A_k^{-1}(y)$ is the time vehicle y passes x_{k+1} minus the time it passes x_k, i.e., the transit time.

The above system differs perhaps from those of the previous sections in that the x_k are arbitrarily selected points from a continuum of possible values. One could imagine placing counters at every point along the highway and observing the time each car passes every point, a complete description of the motion of every car. Instead of considering $A_k(t)$ as counts at discrete locations, we could imagine this to be the values at $x = x_k$ of a function $A(x, t)$.

If we think of $A(x, t)$ as a function on the x, t plane, it can also be drawn as a surface A in a three-dimensional (x, t, n) space as in Fig. 3.2. The curves $A_k(t)$ now represent the lines of intersection of A by planes $x = x_k$, projected onto the (t, n) space, i.e., $A_k(t)$ are contour lines of the surface A drawn on the (t, n) plane.

One can also draw contour maps of the surface A on the other coordinate planes. The curve $A(x, t) = y$ in the (x, t) plane is the contour line of constant cumulative count y. If cars do not pass, the yth cumulative arrival at one location is the same car as the yth

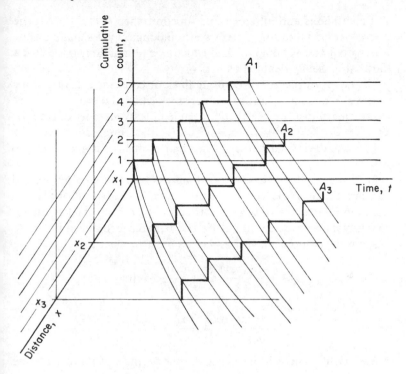

Figure 3.2   *Three-dimensional representation of flow*

cumulative arrival at any other location; thus the curve $A(x, t) = y$ is the 'trajectory' of the yth car. It describes the time at which the jth car reaches x or equivalently the position $x_j(t)$ of the jth car as a function of time.

The contour lines on the (x, n) plane are not used as commonly in practice as the other two. But if one were to take an aerial photograph of the highway (fixed time), one would see a succession of points (cars) on the x-axis (the highway). From this one could draw a graph of the cumulative number of cars from some arbitrary reference point $x = 0$ to a variable x (for each value of t), but, actually, to be consistent with the above diagrams, it is more convenient to count backwards in space, in the opposite direction to the motion. The function $A(x, t)$ for fixed t is the number of cars between the point x and the point further downstream where the reference car labeled 0 is located.

Continuum or fluid approximations are used extensively in the analysis of highway traffic. If we smooth the surface A so as to make it a continuous and differentiable function, then $\partial A(x, t)/\partial t$ is the arrival rate at a location x. Here we should logically designate this by the symbol $\lambda(x, t)$, but in the highway traffic terminology it is called a 'flow' and usually denoted as $q(x, t)$.

In the (x, n) plane, the contour lines of constant t have a slope $\partial A(x, t)/\partial x$. This quantity is negative since the car labels are decreasing with x, but, in absolute value, it is the number of cars per unit length of highway. In the highway traffic teminology, this is the negative of the spatial density of cars and is usually designated as $(-)$ $k(x, t)$.

In the (x, t) plane, the contour lines are trajectories of individual cars and the slope of a contour is the velocity of the car. If one is on a contour line corresponding to $A(x, t) = y$ and one makes an infinitesimal displacement (dx, dt) in the (x, t) plane in such a way as to stay on the contour, then the (dx, dt) must be such that

$$dA = \frac{\partial A(x, t)}{\partial x} dx + \frac{\partial A(x, t)}{\partial t} dt = 0$$

i.e.,

$$\frac{\partial A(x, t)}{\partial x} \frac{dx}{dt} + \frac{\partial A(x, t)}{\partial t} = 0.$$

Since $dx/dt = v(x, t)$ is the velocity at the point (x, t), this can also be written as

$$q(x, t) = v(x, t)k(x, t) \tag{3.1}$$

which is one of the basic identities in continuum traffic flow theory (or fluid mechanics).

Another well-known equation in traffic flow theory derives from the fact that (if there are no exists or entrances) $A(x, t)$ is a single-valued function of x and t and therefore

$$\frac{\partial^2 A(t, x)}{\partial x \partial t} = \frac{\partial^2 A(t, x)}{\partial t \partial x}$$

i.e.,

$$\frac{\partial}{\partial x} q(x, t) + \frac{\partial}{\partial t} k(x, t) = 0 \tag{3.2}$$

or

$$\frac{\partial}{\partial x} [v(x, t)k(x, t)] + \frac{\partial}{\partial t} k(x, t) = 0.$$

This is a familiar form of the conservation equation or the 'equation of continuity'. It describes only one relation between $k(x, t)$ and $q(x, t)$

or $v(x, t)$. To develop any 'theory' this equation must be supplemented with another relation between these observables.

3.5 Tandem queues with finite storage

In most tandem queueing systems, one not only has restrictions on the rate of service, $dD_{ks}(t)/dt \leq \mu_k$ corresponding to a service capacity μ_k of the kth server, but there may also be restrictions on the storage capacity of the 'waiting room' behind the kth server, i.e., one may have constraints of the type

$$Q_k(t) \leq c_k$$

corresponding to a storage capacity of c_k.

Obviously in most situations in which one has such a constraint, the $k-1$th server will be interrupted if $Q_k(t) = c_k$ because the $k-1$th server cannot dispose of a customer if there is no place for him to go. In effect, the queue from the kth server just backs up over the $k-1$th server into the waiting room for $k-1$.

In the highway traffic example with a continuum of service points, one does not have discrete waiting rooms; one has a continuum of storage space. The analogue of the capacity c_k of the waiting room is a restriction on the maximum possible number of cars per unit length of highway, i.e., a restriction on $k(x, t)$. This restriction is usually expressed through a subsidiary condition that $k(x, t)$ is some specified function of $q(x, t)$ or vice versa, but we shall not pursue this further here.

As a first illustration of a system with finite storage, we will assume that customers pass through the system in order, and that a customer who leaves the kth server immediately joins the queue for the $k+1$th server. We also neglect the storage capacity of the server itself, so that

$$D_{kq}(t) \simeq D_{ks}(t) = A_{k+1}(t) \equiv D_k(t).$$

The constraints are

$$0 \leq Q_k(t) = D_{k-1}(t) - D_k(t) \leq c_k, \quad dD_k(t)/dt \leq \mu_k,$$
$$\text{for} \quad k = 1, 2, \ldots, n. \tag{3.3}$$

Instead of having an arbitrarily specified input $D_0(t)$ to the first server, we will imagine that the system is fed through a (perhaps hypothetical) zeroth server of service capacity μ_0 from an infinite reservoir of customers. The zeroth server behaves like any other in the sense that it can be blocked if $Q_1(t) = c_1$. The nth (last) server, in turn, deposits the customers into another infinite reservoir as illustrated schematically in Fig. 3.3(a).

Suppose the system starts at time 0 with arbitrary initial queues,

$$0 < Q_k(0) = D_{k-1}(0) - D_k(0) < c_k, \qquad (3.4)$$

and we wish to follow the future evolution of the $D_k(t)$. Each server

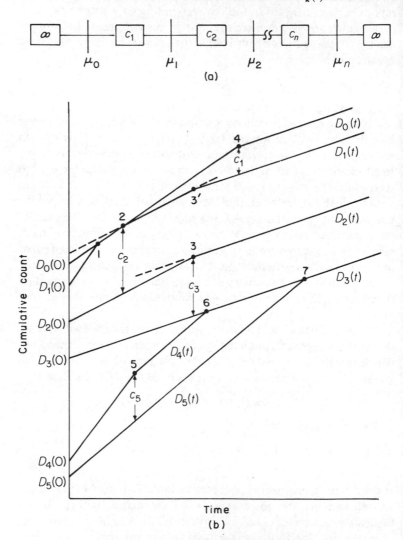

Figure 3.3 *Tandem queues with finite storage*

operates at the fastest rate it can, subject to the constraints implied by (3.3).

One can always construct graphs of $D_k(t)$ iteratively in time. If, as implied by (3.4), we start from some unconstrained state with all storages neither empty nor full, each kth server will start serving at a rate μ_k. Thus we begin by constructing lines

$$D_k(t) = D_k(0) + \mu_k t, \quad k = 0, 1, \ldots, n$$

as in Fig. 3.3(b), until a constraint is violated.

In Fig. 3.3, server 1 has a high service rate, and manages to serve the queue Q_1. After the queue vanishes at point 1, server 1 can serve only as fast as customers arrive; $D_1(t)$ must coincide with $D_0(t)$, at least for a while. Server 2, however, is slower than either 0 or 1. The queue behind server 2 grows until at point 2, the queue has reached the value c_2. The point 2 is identified by translating the curve $D_2(t)$ upward by an amount c_2 as indicated by the broken line, and observing that this curve meets the curve for $D_1(t)$.

Beyond point 2, server 1 can only serve fast enough to keep the storage c_2 full. Since, in this illustration, $\mu_0 > \mu_2$, customers arrive at server 1 faster than it is serving them, and so the queue behind server 1 reforms. One could also say that the queue behind server 2 has backed up over server 1 and that the queue behind server 1 is really the queue for server 2.

Server 3 is slower yet. At point 3, its storage becomes full, i.e., the curve $D_3(t) + c_3$ meets $D_2(t)$. Beyond point 3, $D_2(t)$ we must follow the curve $D_3(t) + c_3$, at a rate μ_3. Since this occurs at a time when the storage c_2 was also full, the slowdown of server 2 also causes a slowdown of server 1 at point 3', i.e., $D_1(t)$ now follows the line $D_3(t) + c_3 + c_2$. In effect, the queue behind server 3 now fills both c_3 and c_2 and starts to fill c_1, which is accomplished at point 4. Beyond point 4, the source is partially blocked; it can serve only at the rate μ_3.

Server 4 is also a fast server and it fills the storage behind server 5 at point 5, but at point 6 it has served all its queue and can serve only at the rate of server 3 thereafter. This causes the full storage behind server 5 to empty at point 7.

In this illustration, the slowest server is server 3. This becomes the ultimate bottleneck; eventually all storages behind server 3 become full, and all storages downstream become empty.

The above example illustrates some of the peculiar things that *can* happen; queues empty and then reform, etc. As a practical matter, however, such things seldom occur because it is unlikely that one would ever find the system in an initial state which would cause them. It is, for example, difficult to imagine how one could generate spontaneously a queue behind server 1 of Fig. 3.3 if this is the fastest server.

There is an alternative way of constructing Fig. 3.3 which tends to produce iteratively various parts of the graph more or less in order of their importance. Consider first the server k_1 with the smallest μ_k (in Fig. 3.3 it is $k_1 = 3$). Clearly this server will never be interrupted either for lack of customers or lack of downstream storage, i.e.,

$$D_{k_1}(t) = D_{k_1}(0) + \mu_{k_1} t \quad \text{for all} \quad t > 0.$$

Consider next the server k_2 with the second smallest μ_k. It can be influenced only by k_1. If $k_2 < k_1$, $D_{k_2}(t)$ will have one and only one slope change, when all storages between k_2 and k_1 are full (server 2 in Fig. 3.3). If $k_2 > k_1$ (as for server 5), $D_{k_2}(t)$ will again have only one slope change, namely when all queues between k_1 and k_2 have disappeared. Actually the servers with $k < k_1$, and with $k > k_1$ behave independently of each other. Thus the slowest server with $k > k_1$ (or $k < k_1$) is influenced only by server k_1 regardless of whether or not it is the second slowest of all servers.

If $k_2 < k_1$, then any server k with $k < k_2$ behaves independently of any server l with $k_2 < l < k_1$ or $l > k_1$. The slowest server in each of these ranges of k has a $D_k(t)$ with at most two slope changes (such as server 4 of Fig. 3.3). The iteration of this is quite straightforward; and since each new curve which is added separates the remaining undetermined curves into two independent groups, the iteration typically goes very quickly. The curves do, however, become more complex with each stage. Server 1 of Fig. 3.3 has four slope changes by virtue of $\mu_1 > \mu_0 > \mu_2 > \mu_3$.

The above graphical construction can also be represented analytically. Clearly, any $D_k(t)$ is either unconstrained or it is constrained by $D_{k-1}(t)$ or $D_{k+1}(t)$, i.e.,

$$D_k(t) = \min\{D_k(0) + \mu_k t, D_{k-1}(t), D_{k+1}(t) + c_{k+1}\}. \tag{3.5}$$

If we iterate this relation, we see that

$$D_k(t) = \min\{D_l(0) + \mu_l t, \quad 0 \le l \le k;$$
$$D_l(0) + \mu_l t + \sum_{m=k+1}^{l} c_m, \quad k+1 \le l \le n\}.$$

Thus $D_k(t)$, if constrained, is either equal to some upstream curve $D_l(t)$, $0 \le l \le k$, with all intermediate storages empty, or it is constrained by some downstream curve $D_l(t)$, $k+1 \le l \le n$, with all intermediate storages full. As illustrated in Fig. 3.3, the relevant constraint will change from time to time, but (3.5) shows that every $D_k(t)$ is eventually determined by the server with the smallest μ_l.

It is interesting (and perhaps useful) to notice that at any time t one can think of the storage c_k as containing a certain number $Q_k(t)$ of occupied spaces and $c_k - Q_k(t)$ unoccupied spaces or holes. Any time a customer passes through server k he leaves a hole at c_k and fills one at c_{k+1}. One could equally well think of the hole as going from c_{k+1} to c_k. Correspondingly one can think of the system shown in Fig. 3.3(a) as a flow of holes from the exist to the entrance via servers n, $n-1, \ldots, 0$.

If at time 0 all storages were full and customer number 0 had just reached the exist, we would say that $D_n(0) = 0, D_{n-1}(0) = c_n, D_{n-2}(0) = c_n + c_{n-1}, \ldots$

$$D_0(0) = c_n + c_{n-1} + \ldots + c_1.$$

If we let $D'_j(t)$ be the cumulative number of holes to have passed the jth server, it is natural to identify the above initial state by $D'_j(0) = 0, j = 0, \ldots, n$, i.e., at $t = 0$ the system has no holes. Subsequently, however, the number of holes to pass server k is the same as the number of cutomers to pass server k, i.e.,

$$D_k(t) - D_k(0) = D'_k(t) - D'_k(0).$$

In general, we have that

$$D'_k(t) = D_k(t) - (c_n + c_{n-1} + \ldots + c_{k+1}). \tag{3.6}$$

One can readily check that the $D'_k(t)$ can be constructed graphically by a procedure analogous to that of Fig. 3.3. When two curves $D'_k(t)$ and $D'_{k-1}(t)$ meet, it represents a full storage or an empty queue of holes, but when $D'_k(t) - D'_{k-1}(t) = c_k$ we have an empty storage or a full storage of holes.

The companion relation to (3.5) is

$$D'_k(t) = \min\{D'_k(0) + \mu_k t, D'_{k-1}(t) + c_k, D'_{k+1}(t)\}.$$

3.6 The effect of finite storage on the capacity of synchronized traffic signals

The analysis of the last section was, in part, a prelude to the treatment of the stochastic behavior of tandem queues. One of the classic design problems for a tandem queueing system is to determine what effect finite storages have on the overall service rate of the system. If there are stochastic effects, it is possible that fluctuations will cause a server upstream or downstream of the bottleneck to serve customers at a rate slower than the bottleneck for a time sufficiently long that the bottleneck is interrupted by a full storage downstream or a lack of customers upstream. The storages c_j should, generally, be chosen so as (usually) to absorb these fluctuations. The bottleneck server should have a storage upstream and an average queue of waiting customers sufficiently large so as to keep busy even during temporary slow periods of the upstream servers, and the downstream servers should have sufficient storage to absorb any queues generated when the bottleneck server temporarily serves customers faster than one downstream.

It is possible also to have deterministic fluctuations in the service rates of successive servers and predictable blocking effects, as illustrated by the following practical situation.

Suppose that highway traffic passes a fixed-cycle traffic signal having a red time r, green time g, and then travels to a second signal also having a red time r and green time g. The trip time of a vehicle between signals is τ, provided it is not delayed by a queue. The off-set, the time difference from the start of green for the first signal until the start of the next green for the second signal is δ, $0 < \delta < r + g$. The flow during green is assumed to be approximately constant whenever there is a queue, having a value μ for both signals. The highway between the signals has an area sufficient to store only c vehicles but is capable of carrying the flow μ, i.e., c is large enough to store vehicles *en route*, $c > \mu\tau$.

We wish to determine the maximum long time average number of vehicles per unit time (capacity) which can travel through the signal pair as a function of δ and the other parameters (c, τ, r, q, μ). We imagine that there is an infinite queue upstream of the first signal and an infinite reservoir downstream of the second signal. There is no turning traffic at either signal.

If the storage c is a limiting constraint, then one would expect the storage to be full when the signal starts green at the second intersection (during every cycle of the signal). We will start the analysis from an initial state with c vehicles in queue and the second signal beginning a green phase.

Consider first the special case in which the first signal starts red at $t = 0$, $\delta = g$. Fig. 3.4 shows a curve $D_1(t)$ for the cumulative departures from signal 1; $A_2(t) = D_1(t - \tau)$, the cumulative number of

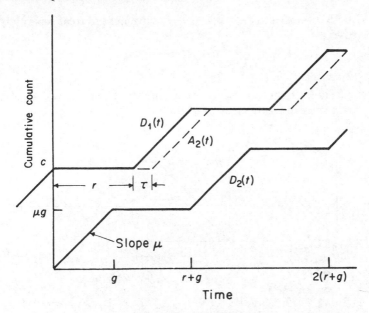

Figure 3.4   *Flow of cars through two signalized intersections*

vehicles which would have reached the second signal in the absence of
a queue; and $D_2(t)$, the number of vehicles to leave the second signal.
The curve $D_1(t)$ remains at c until time $t = r$, and $A_2(t)$ remains at c
until time $r + \tau$. The curve $D_2(t)$ has the form $D_2(t) = \mu t$ until either t
$= g$ or the queue vanishes.

   In order for the queue to vanish during the first cycle, it is necessary
that $g > c/\mu$, i.e., the green time is long enough to serve the stored
queue, and $r + \tau > c/\mu$, i.e., no new vehicles reach the second signal in
time to prevent the queue from vanishing. If *either* $g < c/\mu$ or $r + \tau$
$< c/\mu$, the queue will never vanish, as illustrated in Fig. 3.4. If the
queue does not vanish during the first cycle, it will not vanish in any
later cycles either. Therefore, capacity $= \mu g$ per cycle $= \mu g/(r + g)$ per
unit time.

   The above is true for $\delta = g$, but if the queue does not vanish for
$\delta = g$, it won't for any other δ either. If $g < \delta < r + g$, new vehicles
reach the second signal at time $(r + \tau) - (\delta - g) < r + \tau$, but if $0 < \delta$
$< g$ new vehicles could (in the absence of a queue) reach the second
signal at time τ. In either case, the queue at time g is no smaller than for
$\delta = g$.

If both $g > c/\mu$ and $r + \tau > c/\mu$, it is convenient to treat separately the cases

$$\text{I} \quad c/\mu < g < r + \tau$$
$$\text{II} \quad c/\mu < r + \tau < g.$$

Fig. 3.5 shows some representative geometries of the curves A_1, A_2, D_2 for decreasing sequences of δ values.

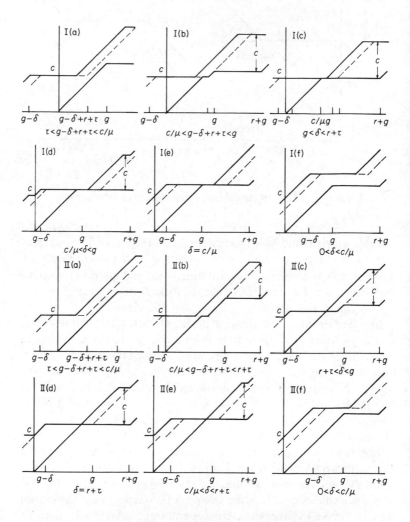

Figure 3.5 *Arrivals and departures from a pair of traffic signals*

For case I(a) with $r < g - \delta + r + \tau < c/\mu$, the curves $A_2(t)$ and $D_2(t)$ do not meet in the first cycle or in any subsequent cycles, so the capacity per cycle is still μg. In I(b), however, for $c/\mu < g - \delta + r + \tau < g$, the queue at the second signal dissipates at time c/μ before the next platoon arrives. The platoon does arrive before the end of the green time but some green time has been wasted. During the red time of the second signal the storage fills again, but this occurs while the first signal is still green. The output from the first signal is blocked until time $r + g$ when the state of the system becomes the same as at $t = 0$. For this range of δ, the capacity per cycle is

$$c + \mu(\delta - r - \tau) < \mu g.$$

In the next range of δ, $g < \delta < r + \tau$, case I(c), the platoon from the first signal arrives after the second signal turns red. Only c cars can be served by the second signal and only c cars can leave the first signal before it is blocked by a full storage. For $c/\mu < \delta < g$, however, case I(d), the first signal is green when the second signal turns green. If the cars leaving the second signal can immediately provide extra storage room for the output from the first signal, then the second signal will be able to serve any cars that could leave the first signal before that signal turned red. For this range of δ, the capacity is

$$c + \mu(g - \delta).$$

For $\delta = c/\mu$, case I(e), the green time is fully utilized and for $0 < \delta < c/\mu$, case I(f), the capacity is back to μg. Fig. 3.6(a) shows the capacity per cycle as a function of δ.

We have assumed here that the area between signals can store c vehicles regardless of whether the vehicles are stopped or moving. In fact, this area cannot hold as many moving vehicles as stopped vehicles and the starting wave generated when the second signal turns green propagates upstream with a finite velocity. The blocking effect, particularly in case I(d), is actually stronger than described here.

In case II, for $r + \tau < g$, II(a) and II(b) are similar to I(a) and I(b). In II(c), however, both signals are green at $t = 0$, as in I(d). If the first signal can keep the queue at c, there will be a flow from the first signal until it turns red. Since the green time is so long, part of the platoon from the second green of signal one can still reach the second signal before it turns red. For $r + \tau < \delta < g$, the capacity has the value $c + \mu(g - r - \tau)$. For $c/\mu < \delta < r + \tau$, case II(e), the capacity increases again, to $c + \mu (g - \delta)$ and then to μg for $0 < \delta < c/\mu$, case II (f). The capacity per cycle is shown in Fig. 3.6(b) as a function of δ.

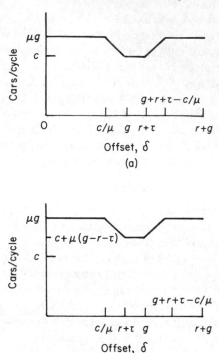

Figure 3.6   *Effect of a storage restriction on the flow through a pair of traffic signals*

3.7   Parallel or multiple-channel servers

For many service systems, a high server capacity is achieved by placing many, m, low capacity single-channel servers in parallel, for example, a toll plaza, grocery store check-out, or telephone exchange. A customer arrival stream is split into component parts according to some rules with each part being served by only one server.

If queues form in such a system, the wait in queue depends upon how the queues are served. At one extreme, there may be separate queues behind each of the m servers, and some rule for partitioning the arrivals, which does not depend upon the queue lengths. The most obvious example is a situation in which there are m types of customers which are (perhaps mistakenly) considered as a single arrival stream, but which, in reality, is the sum of m traffic streams of arrival rate

$$\lambda(t) = \sum_{k=1}^{m} \lambda_k(t).$$

The kth stream desires service only from server k. Thus at a bus depot, one may have m bus stalls (servers) but each bus stall is for a different route. Customers arrive through a common gate, then separate into streams according to their destination, but independently of how long the queues are at the various servers. An airport has a common runway and many gates (servers) but each aircraft is assigned to a particular gate. For such a system, each kth component process serves its arrivals $\lambda_k(t)$, independently of the others, as a single-channel server. Thus we simply have m systems of the type previously analyzed.

At the other extreme, one may have a single queue for all servers; any server which becomes free takes a new customer from the common queue. If the number of servers is small so that, in the fluid approximation, we can neglect the storage capacity of the servers themselves (i.e., the difference between $D_{kq}(t)$ and $D_{ks}(t)$), servers of service rate μ_k would collectively behave like a single server with the service rate

$$\mu = \sum_{j=1}^{m} \mu_k.$$

This has already been implied in some of the previous applications of the fluid approximation.

Even if one had several queues, as at a grocery check-out or a bank, but newly arriving customers join the shortest queue and may also switch queues if they later see a free server, all servers will be kept busy unless all queues vanish. The total queue at all servers will be the same as if there were a common queue. The only difference is that the order of service will, in general, not be FIFO if there are separate queues. Even if customers cannot switch queues but new arrivals always join the shortest queue, the system behaves, in the fluid approximation, as if there were a common queue. In the stochastic version of this, however, one may occasionally have a queue vanish at one server and not another despite the tendency of new arrivals to try to maintain equal queue lengths.

In many practical applications involving multiple-channel servers, however, the number of customers in service is not usually negligible compared with the queue of waiting customers. One of the ad-

vantages of a multiple-channel server is that it may be possible to vary
the number of operating channels according to the demand and
thereby save operating expenses (more tellers, clerks, etc., are
employed during the rush hour than the off-peak). At certain times
one is interested in the queue of customers, at other times in the
number of idle servers.

Although most people picture a multiple-channel server as a
stationary facility consisting of servers which are either idle or busy, a
single bus route with m buses serving the same depot is also an
example of a multiple-channel service. Passengers arrive at a depot
and are willing to use any one of the m buses, whichever one provides
the earliest service. When a bus departs, its service is considered to
begin. It is busy for a certain time, the round trip time (service time),
after which it is available again to load passengers for a new trip. That
the buses actually move is irrelevant to the queue behavior. What the
customer sees is that once a bus has started service, it is unavailable for
a certain period of time, just as in any other multiple-channel service.

This example differs from most of the previous examples of
queueing systems in that a bus will generally serve more than one
customer at a time. In the queueing theory terminology, it would be
called a 'bulk server.' If servers can serve only one customer at a time
(like a taxi), there is no apparent reason why a service should not
commence as soon as there is a free server and a waiting customer.
Indeed, this has been implied in all previous examples without
question. For bulk servers, however, this is not necessarily the best
strategy. Buses are typically dispatched according to a fixed schedule,
or when full, but elevators usually are dispatched as soon as there is a
free elevator and one or more passengers (independent of the load).

Whereas for a multiple-channel server such as a toll collector,
grocery check-out, etc., one would not ordinarily think of the idle
servers as forming a 'queue' waiting for customers, this is obviously an
acceptable notion for a mathematically equivalent system of m taxis
assigned to a cab stand. Idle taxis certainly form queues waiting for
customers.

In Section 3.1, it was stated that one should first identify *all* objects
which satisfy a conservation principle. Taxis obviously do, so do
buses, and so do bank tellers. It is highly advantageous to exploit this
and to view a multiple-channel queueing system as a system with two
types of interacting objects, each of which is conserved. Particularly
for taxis (and consequently for any of the other examples) it is quite
natural that one represents the system schematically as in Fig. 3.7.

Arrivals

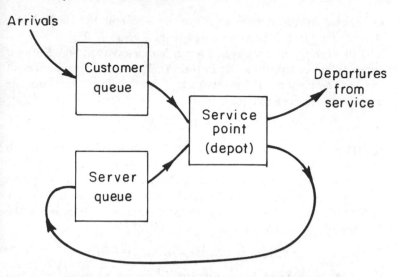

Figure 3.7 *Flowchart for multiple-channel server*

Customers arrive to a queue, enter service and leave service (they could also cycle around and re-enter the system). Servers enter service, but when they complete service they immediately join the queue of idle servers waiting to enter service. The behavior of the system is defined by a specification of conditions under which customers and servers enter service.

Suppose we let $A_c(t)$ denote the cumulative number of customers to arrive at the customer queue, and $D_{qc}(t)$ the number to leave the customer queue and enter service. Let $A_s(t)$ denote the cumulative number of servers to arrive at the server queue, and $D_{qs}(t)$ the number to enter service. As usual

$$Q_c(t) = A_c(t) - D_{qc}(t)$$
$$Q_s(t) = A_s(t) - D_{qs}(t)$$

represent the queue of customers and servers respectively. If both customers and servers are served FIFO in their respective queues,

$$w_c(j) = D_{qc}^{-1}(j) - A_c^{-1}(j)$$
$$w_s(k) = D_{qs}^{-1}(k) - A_s^{-1}(k)$$

are the waiting times or idle times of customer j and server k. The rules of behavior of the system must provide a scheme by which $D_{qc}(t)$, $A_s(t)$ and $D_{qs}(t)$ can be determined from a given $A_c(t)$.

If all service times are equal, a server leaves service at time S after it enters (a bus completes a trip in time S). The cumulative number of service completions $D_{ss}(t)$ by time t is therefore equal to the cumulative number to enter service by time S earlier, i.e.,

$$D_{ss}(t) = D_{qs}(t-S), \tag{3.7}$$

$D_{ss}(t)$ is simply a horizontal translation by S of $D_{qs}(t)$. But each time a service is completed (there is a step in $D_{ss}(t)$), there is also a new arrival to the queue of idle servers, a step in $A_s(t)$. If at time 0 there were no customers and m idle servers, then $Q_s(0) = m$, and $A_s(t) = m$, for $0 < t < S$, until a service has been completed. Thereafter $A_s(t)$ is m plus the number of service completions

$$A_s(t) = m + D_{ss}(t) = m + D_{qs}(t-S) \tag{3.8}$$

i.e., $A_s(t)$ is obtained by translating $D_{qs}(t)$ horizontally by S and vertically by m, as shown in Fig. 3.8(a).

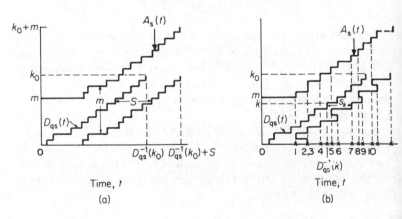

Figure 3.8   *Graphical construction for multiple-channel service*

Even if the service times are not all equal, the values of the service times still determine a relation between the curves $D_{qs}(t)$ and $A_s(t)$. If $D_{qs}^{-1}(k)$ is the time the kth server enters service, it completes service at $D_{qs}^{-1}(k) + S_k$, after a service time S_k. Thus $A_s(t)$ has a step at each of the times $D_{qs}^{-1}(k) + S_k$, $k = 1, 2, \ldots$. From $D_{qs}(t)$ and the S_k one can easily construct the graph $A_s(t)$ as shown in Fig. 3.8(b), but the relation cannot be expressed through a simple formula.

We now have one relation among the four curves $A_c(t)$, $D_{qc}(t)$, $A_s(t)$, and $D_{qs}(t)$. The other relations needed to relate the last three to $A_c(t)$ are determined by the strategy of passing customers and servers past the service point. For most service systems, a customer cannot pass the service point unless a server does also, or vice versa. (An exception to this would be a bus which is dispatched on schedule whether or not there is a passenger.) Thus any steps in $D_{qc}(t)$ and in $D_{qs}(t)$ must occur at the same time. Since neither queue can be negative, these steps can occur only at times when both queues are positive or about to become positive by virtue of a newly arriving customer or service completion.

In general, the construction of $D_{qc}(t)$ and $D_{qs}(t)$ must be done sequentially. If one has constructed the graphs until time t and both queues are nonzero, the strategy will specify (perhaps as a function of the queue lengths and any known future time dependence of $A_c(t)$) whether or not a server will be used, and if so, how many customers it will serve. Certainly, for most efficient strategies, the result of sending customers and servers into service will be that either one or the other or both queues will go to zero. In the case of bulk servers (buses, elevators, etc.) there are many possible strategies. Even though a bus is available and there are passengers waiting, one may choose to delay a dispatch until the bus has some specified number of passengers or until some specified time since the last dispatch. There are certain penalties associated with dispatching a bus with too few passengers. There may also be penalties associated with delaying a dispatch if there will be a high demand for servers at time S later. In any case, a strategy of dispatch will determine both the location and height of the steps of $D_{qc}(t)$ and $D_{qs}(t)$ in terms of the $A_c(t)$ and $A_s(t)$. We will not analyze here the general theory for bulk servers, however.

The special case in which a server serves only one customer at a time is relatively simple. Since each server must enter service with one customer and vice versa, the number of customers who have entered the service must, at all times, be equal to the number of servers that have entered service, i.e.,

$$D_{qc}(t) = D_{qs}(t) \equiv D(t)$$

are the same curve. Furthermore, one would never allow a queue of customers and a queue of servers to exist simultaneously. Therefore either

$$D_{qc}(t) = A_c(t), \quad Q_c(t) = 0$$

or
$$D_{qs}(t) = A_s(t), \quad Q_s(t) = 0.$$

But since $D_{qc}(t) \le A_c(t)$ and $D_{qs}(t) \le A_s(t)$, it follows that
$$D(t) = \min\left[A_c(t), A_s(t)\right]. \tag{3.9}$$

This, along with the previously described relation between $D_{qs}(t)$ and $A_s(t)$ determines both $D(t)$ and $A_s(t)$.

In the case of identical service times, (3.8), $D(t)$ satisfies
$$D(t) = \min\left[A_c(t), m + D(t-S)\right]. \tag{3.10}$$

A graphical solution of this is illustrated in Fig. (3.9) for some arbitrary smooth curve $A_c(t)$ shown by the solid line.

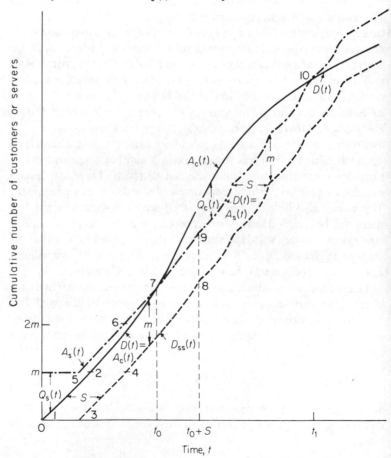

Figure 3.9 *Multiple-channel server with constant service time serving a rush hour*

Starting at time 0 with no customers and m idle servers, i.e., $A_c(0)$ $= 0$ and $A_s(0) = m$, we know that $D(t) = A_c(t)$ at least until such time as $A(t) \geq m$. In Fig. 3.9, knowledge of $D(t) = A_c(t)$ from point 1 to 2 determines $D_{ss}(t) = D(t-S)$ from point 3 to 4 (a horizontal translation by S) and $A_s(t)$ from point 5 to 6 (a vertical translation by m). As long as $A_c(t)$ stays below $A_s(t)$, the curve $A_s(t)$ is determined until time $t + S$ by such horizontal and vertical translations of $A_c(t)$. The vertical distance between $D(t)$ and $A_s(t)$ is the queue of idle servers. At point 7, this queue vanishes, but the known curve of $A_c(t)$ $= D(t)$ to time t_0 determines $D_{ss}(t)$ and $A_s(t)$ until time $t_0 + S$, points 8 and 9 respectively.

While $A_s(t) < A_c(t)$, $D(t) = A_s(t)$ and $A_c(t) - D(t)$ is the queue of customers. Since $D(t) = A_s(t)$ is known from points 7 to 9, this determines $D_{ss}(t)$ and $A_s(t)$ from time $t_0 + S$ to $t_0 + 2S$ and, by iteration, for all later times until the queue vanishes at point 10. During this time, if it lasts more than a time S, $\mu(t) = dD(t)/dt$ is periodic in time with period S (the service rate at time t is the rate at which servers are available which is the rate at which they entered service a time S earlier). The curve $D(t)$ is constructed by repeated translations horizontally by S and vertically by m, of the segment between points 7 and 9.

At point 7 there is a discontinuity in $\mu(t)$ as it switches from $\lambda(t_0)$ to $\lambda(t_0 - S)$ and this discontinuity is repeated at times $t_0 + S$, $t_0 + 2S, \ldots$. At these times, $D(t)$ touches the straight line $D(t_0) + (t - t_0)(m/S)$ corresponding to a constant service rate of m/S. If $\lambda(t)$ is nearly constant for $t_0 - S < t < t_0$, the curve $D(t)$ will stay close to a straight line of slope m/S. Also, point 7 will be nearly the point where the straight line is tangent to $A_c(t)$, as in the basic construction of Fig. 2.3.

After point 10, where $A_s(t) > A_c(t)$, $D(t)$ again follows $A_c(t)$, which in turn generates a $D_{ss}(t)$ and $A_s(t)$ a time S later. If the 'ripple' in the curve $A_s(t)$ generated by the slope change at period 7 is of small amplitude, the queue of customers which vanishes at point 10 is likely to remain zero until there is another rush of arrivals. If, however, the S is sufficiently large or the surge of arrivals at point 7 sufficiently fast, the fluctuating service rate may case a queue to reform. This might happen, for example, with buses or taxis which have a service time S comparable with time required for the rush hour to develop. If $\lambda(t_0 - S)$ is considerably less than $\lambda(t_0)$, there will be a slow rate of returning vehicles immediately after time t_0 but a rapid rate as t approaches $t_0 + S$. This could cause the customer queue to vanish at $t_0 + S$, but a drop in the rate of returning vehicles after time $t_0 + S$

could cause a queue to reform. The queue might vanish again by time $t_0 + 2S$ only to reform. Thus even a 'smooth' $A_c(t)$ can give rise to quite complex queueing patterns.

Fig. 3.9 shows the queue lengths and waiting times for both the servers and the customers. If $A_s(t) > A_c(t) = D(t)$ it shows the server queue (also the customer queue $A_c(t) - D(t)$ which is zero), but, if $A_c(t) > A_s(t) = D(t)$, it shows the customer queue (while the server queue $A_s(t) - D(t)$ is zero). If $A_s(t) > D(t)$, as between time 0 and t_0, the number of idle servers at time t is

$$A_s(t) - D(t) = m + D(t - S) - D(t)$$

the number of busy servers is

$$m - A_s(t) + D(t) = D(t) - D(t - S),$$

i.e., the number of servers used during the previous time interval S. If there is no customer queue at either time t or time $t - S$, then the number of busy servers is $A_c(t) - A_c(t - S)$, the number of arrivals during the time S.

If the arrival rate is nearly constant during the time S, then

$$A_c(t) - A_c(t - S) = \int_{t-S}^{t} \lambda(\tau)d\tau \simeq S\lambda(t). \qquad (3.11)$$

That is not only the number of busy servers but also the number of customers in service. We should recognize that this is a special case of (1.18). Indeed (1.18) suggests that a similar formula would apply even if the service times were unequal, provided we replace S by the average service time.

If $\lambda(t)$ is slowly varying, a more accurate version of (3.11) would be

$$A_c(t) - A_c(t - S) \simeq S\lambda(t - S/2).$$

In effect, the number of busy servers responds to slow changes in $\lambda(t)$ with a time lag of about $S/2$.

3.8 Several customer types

In every queueing situation there is, in effect, a competition among customers for a service which cannot serve everyone at once. In many cases, the customers can be identified or classified in some way. For example, some may be more important than others, or some may have tasks which require less service time. In previous chapters, comparisons between various systems were confined mostly to comparisons of delays for systems with different service rates and consequently also different physical designs. For queues with more than one customer type, however, we have the possibility of using a service of given physical design in many different ways by giving priorities to one customer or another. Much of the literature in queueing theory deals with these priority queues, for the simple reason that it is easier to theorize about the effect of reordering service on an existing facility than about how to design a new service facility.

Some examples in which customers are served in an order other than FIFO are: a grocery store uses an express check-out which, in effect, gives priority service to customers with short service times; an automobile repair shop is also likely to give fast service to customers with small demands; a highway intersection with a traffic signal gives priority to cars from first one traffic stream and then the other; a stop sign gives priority to one stream, in fact most highway maneuvers are governed by a 'right of way' rule; an airport runway serves both landings and take-offs but landings are usually served first; computers usually have a queue of tasks to perform and assign computer time according to some priorities; and any business organization with a backlog of orders will use some priority rule for service.

In the typical queueing problem there is some economic objective (for example, the total cost of queueing delays) which one wishes to minimize. This objective is influenced by the order of service in potentially two ways: first, the time required for the server to perform a set of tasks may be affected by the order in which the tasks are done, and, second, the composition of the queueing and therefore its cost will be affected by the order of service.

Although, in principle, the time required to do some tasks could be a very complicated function of the order of the tasks, in most practical situations the dependence on order is such that the service time of a customer depends only upon the properties of that customer and possibly of those which are served immediately before or after, but not on the long time history of ordering. Frequently a service facility suffers a loss in efficiency if it switches service from one customer type to another. In manufacturing, this loss is sometimes called a set-up time. It may take more time to prepare for a new type task than to continue with a similar task. At a highway intersection controlled by a traffic signal, there is a penalty for switching the signal, a time loss in which the flow of traffic, in effect, drops to zero (yellow signal, start-up time, etc.). Here again, the service is switching from serving customers of one type to those of another type. An airport runway has a service rate of a similar mathematical type except that instead of having a loss associated with a switch in customer type from landing to take-off or vice versa, there is a gain in service rate; the maximum number of movements comes from an alternating pattern. If one is landing both fast and slow aircraft, the rate of service depends also upon the order, it being advantageous to avoid changes in speed.

Section 1.2 shows schemes for drawing and interpreting graphs of arrivals and departures from any service system and for any quantity that is conserved, for example, customers, value of customers, quantity of goods, etc. For systems with several customer types there will usually be many quantities which are simultaneously conserved in the same system. In particular, if customers of a given type cannot change type while in the system, customers of that type are themselves conserved. For customers of type 1, say, one can draw a curve $A_1(t)$ for the cumulative number of arrivals of type 1 customers alone, also a curve $D_1(t)$ for the departures. One can identify on these curves the queue lengths of customers of type 1, delays, etc. The interpretation of such graphs does not depend upon whether or not there are customers of any other type. Each customer type will have its own graphs. The actual shape of the graph, particularly the departure curve, however, will depend upon the times at which customers are served, which in turn will depend upon how the service is shared with customers of other types.

Graphs can also be drawn for the combined arrivals (departures) for any set of customer types; for example, the sum of arrivals (departures) for customers of type 1 plus type 2 in a system with more than two types. One can, of course, also draw graphs for total arrivals (departures) for all types.

If the strategy of service is changed, all departures curves will typically change. In particular, if more service is given to one type of customer thereby increasing the height of its departure curve, the service given to some other customer type will usually decrease. Most questions of optimal strategy are concerned with consequences of trading service between different customer types.

If the cost of delay to a customer of type j is proportional to his delay, then he can be labeled with his cost p_j per unit of delay. These labels are also conserved. If customers of the same type have the same value, then the curves for cumulative arrivals and departures for type j alone, of value or of number, differ only in that the ones for value are scaled by a factor p_j relative to those for number. Different customer types, however, will have different scaling factors. The total number and the total value of all customer types will also be conserved, the curves of cumulative arrivals and departures being obtained by adding the curves for each type. The shape of these combined curves depends upon the 'weights' attached to its component parts. The arrival and departure curves for total number and total value will generally differ by more than just a change of scale.

3.9 Work conserving queues

In order to analyze questions of optimal priorities, it is necessary to specify how a change in ordering of service influences the various departure curves. As noted above, there are situations in which the total time required to serve a specified group of customers depends upon the order in which they are served; there may be penalties associated with a change from service of one type customer to another type. In such cases one must be careful in applying fluid approximations, because the slopes of smoothed departure curves, in which discreteness of the customers is neglected, actually depends upon the details of the sequencing of individual customers. It makes a difference whether a service serving a mix of two type 2 customers for each type 1 customer is done in the order 221221 . . . or in the order 22221122 . . . . The specification of a smoothed departure curve for one type customer does not automatically determine the curve for the second type customer.

There are physical situations, however, in which the total time required to serve any specified group of customers is (nearly) independent of the order in which these customers are served. Such queueing systems are called 'work conserving', work being identified as the amount of service time.

Since for a work conserving service, the service time of a customer is not affected by the order in which the customers are served, an order which might depend upon who else is in the queue when the customer arrives or even who arrives later, one might imagine that his service time is specified when he joins the queue or even before. Even if the service time is random and is, in fact, determined only when the service is completed, one could imagine a hypothetical random sampling of the service time which is performed earlier. The service time or work can be considered a label which the customer brings into the queue and retains throughout his stay in the system. The cumulative amount of work brought to the queue by any class of customers can be treated in the same way one would treat any other conserved quantity, i.e., one can draw curves like Figs. 1.3 or 1.4.

To analyze work conserving queues, it is advantageous to consider first the cumulative arrival and departure curves for the total work of all customers. If $A^w(t)$, $D_q^w(t)$, and $D_s^w(t)$ denote the cumulative amount of work to arrive, leave the queue, and leave the service, respectively, then each of these are step functions with each step of height equal to the service time of the customer which generated in step.

If the server is a single-channel server, then each time a customer of service time S_k enters the server causing a step of height S_k in $D_q^w(t)$ at time t_{qk}, the server will be busy until time $t_{sk} = t_{qk} + S_k$. If another customer is waiting at time t_{sk}, he will enter service at time t_{sk}. Thus a vertical step of height S_k in $D_q^w(t)$ is followed by a horizontal step of width S_k as shown in Fig. 3.10. Correspondingly at time t_{qk}, the curve $D_s^w(t)$ starts a horizontal leg of length S_k to be followed by a vertical step of height S_k at time t_{sk}.

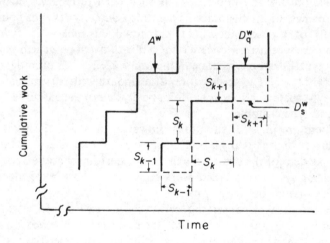

Figure 3.10 *Departure curves of work from a single server*

The most important feature of the curves $D_q^w(t)$ and $D_s^w(t)$ is that, as long as there is a queue, these curves are both step function approximations to a line of slope 1 (with a maximum deviation of S_k), even though the steps S_k may be unequal. A fluid approximation to the work departure curves would disregard such details and yield a smooth curve independent of the order in which customers are served.

For an m-channel server, the departure curve is the sum of the departure curves for the individual servers. If we can still disregard irregularities in the $D_q^w(t)$ caused by the individual steps and also clustering of the S_k from different servers (at most the sum of m service times), then the curve $D_q^w(t)$ will stay close to a straight line of slope m as long as there is a customer queue (independent of the queue discipline). If m is sufficiently large, we will, of course, still have the type of phenomena discussed in Section 3.6 which is associated with clustering of the service starts of different servers.

To illustrate the use of fluid approximations in the comparison of queue disciplines, suppose we have only two customer types labeled 1 and 2, and a single-channel work conserving server. The service times of the two customer types may be different. Let

$$s_j = \text{mean service time for customers of type } j.$$

If $A_j(t)$ is the fluid approximation to the number of j type customers to arrive by time t, then $s_j A_j(t)$ represents the cumulative work to arrive by time t for type j customers. The total work to arrive by time t is

$$s_1 A_1(t) + s_2 A_2(t).$$

We can immediately construct graphs of cumulative arriving and departing work as shown in Fig. 3.11. These are drawn in the same way as Fig. 2.3 except that the vertical scale is work instead of customer number and the departure curve has maximum slope 1. Quite independent of the order of service, the vertical distance between the arrival and departure curves gives the work in the queue. The graphs also shows the busy period, the times when there is a nonzero queue.

We can also draw on the same graph the cumulative arriving work for customers of type 1 and 2 separately, the sum of which is the curve $s_1 A_1(t) + s_2 A_2(t)$. We have not yet specified the rate of departure for work of type 1 or 2 separately, but regardless of how the rate of service is partitioned between the two customer types, the sum of the work rates must give the departure curve of Fig. 3.11. Also, the work in the queues must sum to that shown for the combined system, and the areas between the component arrival work and departure work curves must sum to that for the combined system.

If, in addition to the above information, we also specified the curve $D_1(t)$, or $s_1 D_1(t)$ we could immediately evaluate everything else of interest. The work in queue for type 1 customers is $s_1 A_1(t) - s_1 D_1(t)$.

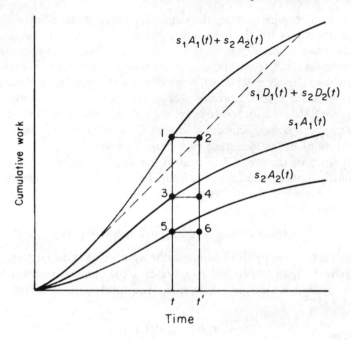

Figure 3.11 *Cumulative work for a work conserving server*

If we subtract $s_1 D_1(t)$ from the curve for total work departure, we obtain $s_2 D_2(t)$, and from this the work in queue for type 2 customers. The curves $A_1(t)$, $A_2(t)$, $D_1(t)$ and $D_2(t)$ for the cumulative numbers of customers are obtained by rescaling each of the work curves by s_1 or s_2. The total number of arrivals and departures are then found by addition of these. The curve for total departures $D_1(t) + D_2(t)$ will not generally have a constant slope over the busy period, however, as did the curve for total work.

If the cost per unit of delay of customers of type j is p_j, we can also draw curves for $p_j A_j(t)$ and $p_j D_j(t)$ by scaling the curves of numbers of customers by p_j, or the work curves $s_j A_j(t)$ by a factor p_j/s_j. Curves for combined costs are then obtained from addition of these component cost curves.

There are many types of queue disciplines that one could employ, different ones being optimal with respect to different objectives. We consider below two simple examples.

(a) *Priority service*

If customers of type 1 always obtain service ahead of customers of type 2, then the departure curve of type 1 customers is that which would exist if there were no type 2 customers. In Fig. 3.11 we can immediately construct the curve $s_1 D_1(t)$ by drawing a curve of maximum slope 1 below the arrival curve $s_1 A_1(t)$ in the same way that the composite departure curve was constructed from the composite arrival curve. The departure curve $s_2 D_2(t)$ is now obtained by subtraction of $s_1 D_1(t)$ from the composite departure curve.

If the cost of the queues is p_j per unit of delay for type j customers, the total cost of the jth queue is p_j/s_j times the area between the $s_j A_j(t)$ and $s_j D_j(t)$ curves. These component areas represent a partition of the area between the composite arrival and departures curves, the latter being independent of queue discipline. If $p_1/s_1 > p_2/s_2$, the strategy which minimizes total cost is clearly that strategy (i.e., that partition of the total area) which gives the smallest possible area to the type 1 customers, subject to the constraint that the total work rate and the work rate of each component must be between 0 and 1. The priority service is the optimal work conserving strategy with this cost structure.

(b) *First in, first out*

The feature which characterizes FIFO service is that the delay suffered by a customer who arrives at time t is independent of which type of customer it may be. The delay is the amount of work in the queue at time t.

We can obtain the work departure curves $s_1 D_1(t)$ and $s_2 D_2(t)$ as follows. First, draw the given curves $s_1 A_1(t), s_2 A_2(t)$, and $s_1 A_1(t) + s_2 A_2(t)$; then construct the combined departure curves $s_1 D_1(t) + s_2 D_2(t)$ with maximum slope of 1 as in Fig. 3.11. For any value of t draw a vertical line up to the composite arrival curve (point 1) of Fig. 3.11, a horizontal line over to the departure curve at point 2, and then a vertical line back down to the axis at time t'. The horizontal distance $t' - t$ between the two vertical lines is the delay to any customer who arrives at time t. The cumulative work of type 1 arrivals at time $t, s_1 A_1(t)$ is represented by the height at point 3. A type 1 customer who arrives at time t, must depart at time t', therefore point 4 at height $s_1 A_1(t)$ must lie on the departure curve $s_1 D_1(t)$ indicating that, at time t', the cumulative number of departures is equal to the cumulative number of arrivals to time t. Similarly, point 6 at height $s_2 A_2(t)$ must be on the departure curve $s_2 D_2(t)$. The locus of points such as 4 and 6 generated from all values of t must be the curves $s_1 D_1(t)$ and $s_2 D_2(t)$.

3.10   Queueing at freeway ramps

To illustrate how some of the methods of the previous sections can be combined, we consider the queueing which occurs at a freeway off-ramp followed by an on-ramp (a typical geometry for a 'diamond interchange') as shown schematically in Fig. 3.12(a). Suppose that,

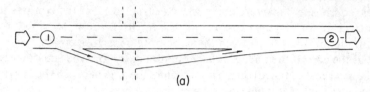

(a)

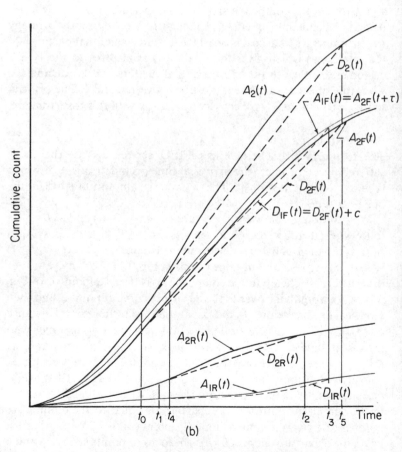

(b)

Figure 3.12   *Graphical evaluation of queues at freeway ramps*

during the rush hour, the on-ramp merge, point 2, becomes a bottleneck causing a queue to form on the freeway and perhaps also on the ramp. When this queue backs up past the off-ramps, point 1, it interferes with the traffic which wishes to exit at 1. We are particularly interested in the 'secondary' queueing, the delay caused by the bottleneck at point 2 to the off-ramp vehicles at point 1 which do *not* traverse the bottleneck.

The details of the merge near point 2 are not of interest, except insofar as the geometry of the merge section may influence a priority at the merge, which, in turn, will determine how many vehicles queue on the freeway and how many queue on the ramp. For any well designed merge section, it is not the merge area itself that is in the bottleneck, it is the freeway immediately downstream from the merge. The capacity of the bottleneck is, therefore, independent of how the service is divided between the main stream and the ramp. Thus point 2 acts like a work conserving system relative to the through traffic and the ramp traffic. The combined delay to ramp traffic plus through traffic past 2 is, therefore, independent of the priority rule at the merge.

Let $A_{2R}(t)$ and $A_{2F}(t)$ represent the cumulative number of vehicles which, in the absence of queueing, would have passed point 2 by time t from the ramp and from the freeway respectively, and let

$$A_2(t) = A_{2R}(t) + A_{2F}(t).$$

If the capacity service rate of the merge is μ, the 'queue' of vehicles at point 2 can be found by the usual graphical construction of Fig. 2.3, as shown also in Fig. 3.12(b). The queue is defined here as the number of vehicles which wished to pass point 2, less those which actually did. This is not the same as the number of vehicles in the 'physical queue'. The latter includes the former plus all vehicles which, in the absence of queueing, would have been in the stretch of highway which is overrun by the queue.

The area between $A_2(t)$ and $D_2(t)$, however, represents the total delay to all vehicles which pass point 2. This area is independent of the priority rule at the merge, whether the vehicles queue on the freeway or the ramp, the lengths of the physical queues as compared with those defined here, or any other physical characteristics of the flow (densities, velocities, etc.) behind the bottleneck. Since the off-ramp at point 1 is assumed not to be itself a bottleneck, it will not prevent vehicles from passing point 2 at a rate μ while there is a queue. The delays at 2, as described above, are also independent of what happens to the off-ramp traffic at 1.

The above illustrates how, in constructing models, one should not only recognize what data are relevant to the questions being posed, but also what is *irrelevant*. In analyzing some physical problem one certainly does not wish to collect any more input data than is necessary, or to make any more postulates than necessary about things one cannot measure.

Although the priority rule at 2 has no effect upon the total delay to the traffic passing 2, it does affect how the queue is partitioned between the freeway and the ramp, which, in turn, will determine whether or not the freeway queue backs up to block the off-ramp at 1.

For most high-speed on-ramps, merge sections are designed such that, if a queue were to form on both the freeway and the ramp, vehicles from the ramp and the outer freeway lane would mix in approximately equal portions (alternate). Under such conditions, the merge acts like two independent queueing systems, each with a fixed service rate, the ramp having an effective service rate μ_r equal to about half the maximum flow of a single freeway lane (μ_r is about 1000 vehicles per hour), the freeway having a service rate for through traffic of $\mu - \mu_r$.

If, however, the ramp flow is less than μ_r and there is no queue on the ramp, the merge acts like a priority service giving priority to the ramp. The ramp queue remains zero, but the through traffic can use all the excess service. Correspondingly, if the freeway through traffic flow is less than $\mu - \mu_r$ and the queue on the freeway is zero, the freeway obtains priority service. Any unused service can be used by the ramp traffic (up to some other limit, perhaps $2\mu_r$, or the capacity of ramp itself).

In principle, the strategy of service could switch many times during a rush hour among the three cases above and yield rather complex departure curves, but, in practical applications, this is quite unlikely. The most common situation is that for which the ramp has priority service and the queue forms only on the freeway. Thus in Fig. 3.12(b) the difference $A_2(t) - D_2(t)$ represents the queue on the freeway.

Before analyzing the behavior at point 1, it will be necessary to draw the curves for $A_{2F}(t)$ and $D_{2F}(t)$. Fig. 3.12(b) illustrates a hypothetical situation in which the queueing starts on the freeway at time t_0, i.e., $\lambda_{2R}(t_0) < \mu_r$. During a later time, however, $t_1 < t < t_2, \lambda_{2R}(t) > \mu_r$.

The graphical procedure is first to treat the ramp traffic as if it had a separate server of service rate μ_r, and construct $D_{2R}(t)$ in the usual way from $A_{2R}(t)$. From the curves $A_2(t)$ and $D_2(t)$ obtained in the same way but with service rate μ, evaluate the difference

$$D_{2F}(t) = D_2(t) - D_{2R}(t).$$

This can be done directly from the graphs. For any value of t, one measures $D_{2R}(t)$ with dividers and subtracts it from the height $D_2(t)$.

The $A_{2F}(t)$ represents the number of through vehicles which, in the absence of queueing, would have reached point 2 by time t. Let $A_{1F}(t)$ represent the number which would have reached point 1 by time t. If the transit time, in the absence of queueing, between points 1 and 2 is τ, then

$$A_{1F}(t) = A_{2F}(t + \tau).$$

Although τ does vary somewhat with the flow, we will consider it to be a given constant. Thus $A_{1F}(t)$ is simply a horizontal translation of $A_{2F}(t)$.

If the physical queue behind point 2 has not reached point 1, $A_{1F}(t) - D_{2F}(t)$, the number of vehicles which have passed point 1 but not point 2, represents the number of vehicles between points 1 and 2 (either moving freely or in queue). Suppose the storage capacity c of the highway section between points 1 and 2 is known either by calculations from the geometry of the highway and the density of queueing vehicles, or by direct observation. Actually the value of c will also vary somewhat with the flow, $dD_{2F}(t)/dt$, but we will assume that it is a constant.

The queue will back up to point 1 when $A_{1F}(t)$ is equal to $D_{2F}(t) + c$. If $A_{1F}(t) > D_{2F}(t) + c$, the number of through vehicles which have passed 1 is

$$D_{1F}(t) = D_{2F}(t) + c,$$

a vertical translation of $D_{2F}(t)$. The difference $A_{1F}(t) - D_{1F}(t)$, of course, represents the queue of through vehicles behind point 1. Again, this refers to the number of vehicles which in the absence of a queue would have reached point 1 but which have been delayed; it is not the number of cars in the physical queue. The c, however, which measures the number of vehicles between points 1 and 2 does relate to the physical queue.

Fig. 3.12(b) obviously was not drawn so as to minimize the number of crossing curves and thereby present the most elegant schematic picture. Neither has it been drawn from actual data, but it is intended to show something which is reasonable. If the vertical range of the graph were about 10 000 vehicles and the horizontal range about 1 hour, the maximum queue length in the figure is about 600 vehicles on the freeway (a physical queue of about 700) and 100 on the ramp. The transit time is about one minute and c about 500.

Finally, we are ready to consider the delay to the ramp traffic at 1 for some given cumulative curve $A_{1R}(t)$ for the off-ramp.

If it were possible to provide priority service to the ramp traffic at point 1 (a by-pass of the queue), the delay to the ramp traffic would be zero. There would be no penalty to the through traffic as long as the through vehicles could pass point 1 at a rate sufficient to prevent point 1 from becoming the bottleneck. If the freeway had m lanes and the queue could be stored on $m-1$ lanes, this would clearly increase the physical length of the queue but it would not increase the delay; the queue would move faster.

Unfortunately it is not usually possible to provide priority service without building extra traffic lanes. If there are m traffic lanes at point 2, the on-ramp is using at most the equivalent of half a lane; therefore, the through traffic is taking at least $m-1/2$ lanes. If there are also m lanes at point 1, the through traffic cannot be squeezed onto $m-1$ lanes without causing a bottleneck. Thus the outer lane at 1 must contain a mix of through traffic and ramp traffic.

Even though some traffic lanes may move faster than others for a short time, drivers tend to crowd fast lanes and force them all to move at approximately the same speed. Drivers tend to believe in FIFO and try to achieve such. It is, therefore, reasonable to assume that the queue discipline past point 1 is FIFO for the combined traffic $A_{1F}(t)$ and $A_{1R}(t)$.

With FIFO discipline for both $A_{1F}(t)$ and $A_{1R}(t)$, the construction of $D_{1R}(t)$ follows the same procedure as in Fig. 3.11. A ramp vehicle which arrives at 1 at time t suffers the same delay as a through vehicle arriving at time t, as measured by the horizontal distance $D_{1F}^{-1}(k)$ $- A_{1F}^{-1}(k)$. The queue for the ramp exists for the same time interval (t_4 to t_5 in Fig. 3.12(b)) as the queue of through traffic behind point 1.

The scale of Fig. 3.12(b) is not designed to show the ramp queue at 1 with high accuracy. One could magnify the scale for the ramp traffic to achieve better accuracy, but if one is interested in the total delay for all queues, one need not know the ramp queues to a smaller absolute error than that of the through traffic. For such purposes it is appropriate to use a common scale for all queues regardless of their relative sizes.

3.11 Nonlinear cost of delay

Although it is usually assumed that queueing costs are proportional to the total delay, at least within customer classes, this is seldom really true. If one uses a priority discipline and the high priority customers arrive at a rate larger than μ, so as to saturate the service, the lower priority customers receive no service until the queue of high priority customers vanishes. Although this may minimize total cost, under the hypothesis that costs are proportional to delays, this cost may be carried mostly by only a few customers who suffer very long delays. In many practical situations this is not considered a desirable consequence. There must, therefore, be an implied price associated with excessive delay to any individual customer.

Some of the above difficulties can be avoided by introduction of more realistic costs. This is seldom done in practice, however, because one usually does not know what constitutes a reasonable cost and the introduction of delay-dependent costs introduces some mathematical complications. Sometimes various artificial constraints are introduced. For example, one could retain the usual price structure but minimize 'cost' subject to the constraints that the difference or the ratio of queue lengths of two customer types must never exceed some preassigned value.

In addition to these cases in which the assumption of a cost proportional to delay leads to 'optimal' strategies which are not consistent with accepted practice, there are other cases in which the cost of a delay is clearly not proportional to the delay. If the queue is a storage of perishable goods waiting to be processed, the cost of long delays is disproportionately higher than for short delays. For short delays the cost involves only storage costs or the cost of delay in not having a finished product, but long delays may result in an inferior product. In other cases, long delays to people may result in missed appoints; long delays to landing aircraft may involve risk of accidents, etc.

In most situations it is still reasonable to assume that the total cost of all delays is the sum of the costs of delays to all customers, but the cost of a delay to an individual customer is not necessarily proportional to his delay. Suppose, instead, that each increment of delay to a jth customer who has already been delayed a time w is $p_j(w)$ times the increment of delay and that $p_j(w)$ is an increasing function of w. The total cost of delay w to this customer is $P_j(w)$,

$$P_j(w) = \int_0^w p_j(t)\,\mathrm{d}t, \quad p_j(w) = \mathrm{d}P_j(w)/\mathrm{d}w. \quad (3.12)$$

If customers can be classified into type (two types for example) such that all customers of the same type are identical, then each customer type will have a single cost function, $p_1(w)$ or $p_2(w)$. We consider separately the cumulative arrival and departure curves for each type. If the departure curve for customers of type j is independent of the queue discipline among customers to type j, the optimal strategy is to serve customers of type j in the order FIFO (see problem 1.3).

Some of the interpretations of Sections 1.2 and 1.4 can be generalized to treat a nonlinear cost function. For customers of type 1, say, with FIFO order within this type, the delay to the kth type 1 arrival is the horizontal distance $w_1(x)$ between $A_1(t)$ and $D_1(t)$ at height x for $k-1 < x < k$.

$$w_1(x) = D_1^{-1}(x) - A_1^{-1}(x).$$

The cost to this customer is $P_1(w_1(x))$. It is also equal to $P_1(w_1(x))$ times the vertical segment between $k-1$ and k (i.e., times one), or, since $w_1(x)$ is independent of x for $k-1 < x < k$,

$$\text{cost to customer } k = \int_{k-1}^{k} P_1(w_1(x))\,dx = \int_{k-1}^{k}\int_{0}^{w_1(x)} p_1(\tau)\,d\tau\,dx$$

$$= \int_{k-1}^{k}\int_{A^{-1}(x)}^{D^{-1}(x)} p_1(t - A^{-1}(x))\,dt\,dx. \qquad (3.13)$$

Although it may appear that we are artificially making things more complicated by replacing a function $P_1(w_1(x))$ by a double integral, (3.13) does have some advantages. We can think of $dt\,dx$ as an element of area in the (x, t) plane of Fig. 3.13 and (3.13) as the integral of the marginal cost $p_1(t - A^{-1}(x))$ over an area between $A_1(t)$ and $D_1(t)$ and between heights $k-1$ and k. The total cost to all customers of type 1 who arrive during a busy period, i.e., between times when the queue vanishes, is the sum of (3.13) over k. The sum of the integrals over all horizontal strips $k-1 < x < k$, however, is equal to the integral over the entire region R_1 between the curves $A_1(t)$ and $D_1(t)$,

$$(\text{total cost})_1 = \iint_{R_1} p_1(t - A_1^{-1}(x))\,dx\,dt. \qquad (3.14)$$

The step functions $A_1(t)$ and $D_1(t)$ can now be approximated by a smooth curve (fluid approximation). To evaluate costs from graphs of cumulative work, one need only observe that for customers of type 1, the work curves and count curves differ only by a scaling of the x

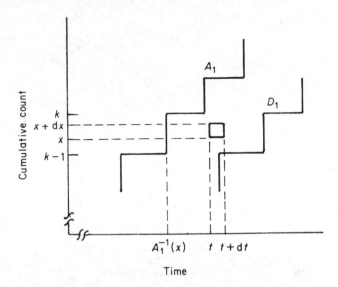

Figure 3.13   *Representation of an element of area between curves A_1 and D_1*

coordinate by s_1. If we let $y = s_1 x$, (3.14) transforms to

$$(\text{total cost})_1 = (1/s_1) \iint\limits_{R_1'} p_1(t - A_1^{-1}(y/s_1)) \, dy \, dt, \qquad (3.15)$$

with R_1' the region between the work curves $s_1 A_1(t)$ and $s_1 D_1(t)$.

In the special case in which p_1 is independent of w, the integrals (3.14) and (3.15) become p_1 times the areas of R_1 or R_1' as discussed previously.

The integral (3.14) or (3.15) can also be interpreted as the volume under a surface in a three-dimensional (t, x, p) space. This surface can be constructed directly from graphs of $A_1(t)$, $D_1(t)$, and $p_1(w)$ as illustrated in Fig. 3.14. Cut a template of the shaded area for the curve $p_1(w)$ and place it perpendicular to the (t, x) plane with the w-axis of the template in the (t, x) plane parallel with the t-axis. Now let the origin of the template travel along the curve $A_1(t)$. The surface generated by the curve $p_1(w)$ is $p_1(t - A^{-1}(x))$.

One can also draw the contour map of $p_1(t - A_1^{-1}(x))$ by translating the curve $A_1(t)$ in the direction of the t-axis; a translation by w will produce the contour with value $p_1(w)$. To draw these contours, make a template of $A_1(t)$ and translate the template.

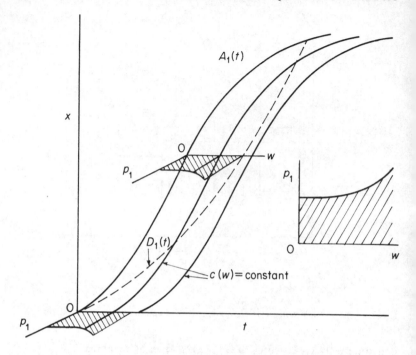

Figure 3.14   *Evaluation of total cost*

In a typical optimization problem with two customer types, one
might specify the arrival curves $A_1(t)$ and $A_2(t)$, the marginal cost
functions $p_1(w)$ and $p_2(w)$, and one relation between $D_1(t)$ and $D_2(t)$.
If, for example, the service is work conserving, this last relation would
state that $s_1 D_1(t)$ and $s_2 D_2(t)$ must sum to the curve for departure of
total work as constructed in Fig. 3.11. The problem is to determine
$D_1(t)$, and consequently $D_2(t)$, so as to minimize the total cost for both
customer types, the sum of two expressions like (3.14) or (3.15). If
$p_1(w)$ and $p_2(w)$ were constants, the optimal solution would give
priority service to the customer with the larger p_j/s_j as shown in
Section 3.9, but if $p_1(w)$ and $p_2(w)$ are not constant, any fixed priority
service might give very large delays to the low priority customer,
which could be very costly.

If we translate the time origin so that $t = 0$ is the time when a queue
first forms, as in Fig. 3.11, and we count arrivals and departures from
this time origin, then the departure curve for total work becomes

$$s_1 D_1(t) + s_2 D_2(t) = t \qquad (3.16)$$

at least until such time when $s_1 A_1(t) + s_2 A_2(t) = t$; the queue vanishes again.

Any allowed strategy for service is equivalent to a specification of a curve $D_1(t)$ or $s_1 D_1(t)$ subject only to the restrictions

$$s_1 D_1(t) \leq s_1 A_1(t), \quad s_2 D_2(t) \leq s_2 A_2(t), \qquad (3.16a)$$

i.e., both queues are non-negative, and

$$0 \leq s_1 \, dD_1(t)/dt \leq 1, \qquad (3.16b)$$

i.e., the rate at which work is done for customers of type j must be non-negative, which with (3.16) implies that it is also less than 1. Thus our problem is to find among all curves $s_1 D_1(t)$ satisfying (3.16a, b), with $s_2 D_2(t)$ given by (3.16), that curve which minimizes the total cost.

To compare strategies, we first observe that the value of p_j $(t - A_j^{-1}(y/s_j))$ at any point (t, y) depends upon the curve $A_j(t)$, but not $D_j(t)$. The total cost for customer type j depends upon the curve $D_j(t)$ only because $D_j(t)$ determines the boundary of the region of integration in (3.14) or (3.15).

For any proposed optimal curves $s_1 D_1(t)$ and $s_2 D_2(t)$ satisfying (3.16), we consider also neighboring curves

$$s_1 D_1(t) + \delta(t) \quad \text{and} \quad s_2 D_2(t) - \delta(t) \qquad (3.17)$$

which also satisfy (3.16). Suppose that $\delta(t)$ is uniformly small for all t and that (3.17) also satisfy (3.16a, b). The change $\delta(t)$ in $s_1 D_1(t)$ induces a change in cost to customers of type 1 equal to the integral of $(1/s_1)p_1(t - A_1^{-1}(y/s_1))$ over a strip between the curves $s_1 D_1(t)$ and $s_1 D_1(t) + \delta(t)$, i.e., the difference between two integrals like (3.15). Similarly, it induces a change in the cost to customers of type 2 equal to the integral of $(1/s_2)p_2(t - A_2^{-1}(y/s_2))$ over a strip between the curves $s_2 D_2(t)$ and $s_2 D_2(t) - \delta(t)$. The net change in cost will be

$$- \int \delta(t)(1/s_1)p_1(t - A_1^{-1}[D_1(t)])dt$$
$$+ \int \delta(t)(1/s_2)p_2(t - A_2^{-1}[D_2(t)])dt$$

integrated over the range of t when the queue is nonzero. These two integrals can be combined into a single integral

$$\int \delta(t)[-(1/s_1)p_1(t - A_1^{-1}[D_1(t)])$$
$$+ (1/s_2)p_2(t - A_2^{-1}[D_2(t)])]dt. \qquad (3.18)$$

The interpretation of $p_j(t - A_j^{-1}[D_j(t)])$ is that it represents the marginal cost per unit delay to the customer of type j who, under the prevailing strategy with departure curve $D_j(t)$, is to be served at time t, under FIFO discipline among customers of the same type. The quantity in square brackets can be interpreted as the difference in the rates at which marginal costs will be served for the two customer types at time t under the existing strategy.

The cost differential (3.18) can be used to establish necessary conditions for an optimal solution. If $D_1(t)$ is an optimal solution, then (3.18) must be non-negative for *all* admissible $\delta(t)$, for otherwise there would be a curve $s_1 D_1(t) + \delta(t)$ which gave less cost than $s_1 D_1(t)$.

If there exists some range of time from α_1 to α_2 over which the optimal $D_1(t)$ satisfies strict inequalities in (3.16a, b), then, over this range of time, $\delta(t)$ can be either positive or negative and could have either a positive or a negative derivative. If the bracket in (3.18) were not identically zero over the time α_1 to α_2, there would be an admissible $\delta(t)$ which is negative at any time when the bracket is positive and vice versa, making the integrand and the integral (3.18) negative. This would contradict the assumption that $D_1(t)$ is optimal. Over this range of t, the bracket must, therefore, be identically zero.

From this we conclude that the optimal $D_1(t)$ must satisfy one of the following conditions at every value of t:

$$D_1(t) = A_1(t) \quad \text{queue 1 vanishes} \qquad (3.19a)$$
$$D_2(t) = A_2(t) \quad \text{queue 2 vanishes} \qquad (b)$$
$$s_1 dD_1(t)/dt = 1 \quad \text{no service for customers of type 2} \qquad (c)$$
$$s_1 dD_1(t)/dt = 0 \quad \text{no service for customers of type 1} \qquad (d)$$

or

$$(1/s_1)p_1(t - A_1^{-1}[D_1(t)]) = (1/s_2)p_2(t - A_2^{-1}[D_2(t)]), \qquad (e)$$

the marginal rates of serving costs are equal for the two customers. We still have the problem of determining which of these conditions apply at any time. In principle this can still lead to a very difficult problem because, for sufficiently complex $A_j(t)$ or $p_j(w)$, the choice among these possibilities may change many times, but in most practical situations the optimal strategy will be fairly simple. Whereas (3.19) represents a necessary condition, if (3.19e) is not satisfied at any time because $p_1/s_1 > p_2/s_2$ for example, then there must still be some constraint arising from (3.16a, b) that prevents $\delta(t)$ from being positive; there is something that prevents giving additional service to type 1 customers.

The optimal strategy will usually (but not always) be one in which priority service is given to customers of type 1 whenever $p_1/s_1 > p_2/s_2$, or to type 2 whenever $p_1/s_1 < p_2/s_2$, but if $p_1/s_1 = p_2/s_2$, the service should be shared in such a way as to preserve this condition (if possible). If, for example, with zero queues at time 0, $p_1/s_1 > p_2/s_2$ one will start by giving priority service to type 1. As a result of this strategy, however, the queue of type 2 customers will grow and p_2/s_2, will increase. If at some time p_2/s_2 should become as large as p_1/s_1, the two customer types become, in essence, equivalent and one will hereafter try to keep them that way. If one should make a mistake and continue the priority to type 1 too long, one would then find $p_1/s_1 < p_2/s_2$ and reverse the priorities. This reversal would tend to bring the cost rates back into balance again. Thus the strategy of maintaining $p_1/s_1 = p_2/s_2$ is stable.

In the above strategy, the action one takes at time t depends only upon the state at time t, in particular the values of p_1/s_1 and p_2/s_2 for customers that are candidates for the service at that time. If this is optimal, it would imply that the overall optimal policy is to do at each time t what is 'locally optimal'.

Exceptions to the above do occur. Suppose, for example, that type 2 customers arrive so rapidly that when p_2/s_2 becomes equal to p_1/s_1, it is not possible to serve customers of type 2 fast enough (even with priority service) to prevent p_2/s_2 from overshooting p_1/s_1. In such cases the optimal strategy would anticipate the approaching dominance of type 2 and start priority service to type 2 before p_2/s_2 equals p_1/s_1. One would start to reduce the queue of type 2 customers to get ready for the surge. We shall not pursue this further, however, because there are many such exceptions which on the one hand are somewhat tedious to analyze and on the other hand are not likely to occur very frequently in practical problems.

3.12 A baggage claim

Some queueing systems have special rules for service quite different from any of the classic examples described above. As an example, consider the queue of passengers from an airplane who wish to claim their luggage from a baggage carousel.

Passengers leave an aircraft and walk to the baggage claim area. The bags are removed from the aircraft and transported to a conveyor belt (or other device) where they are handled at some finite rate; they are not all available simultaneously.

Suppose there are N_b bags and N_p passengers having one or more bags.

Let

$N_b F_b(t)$ = cumulative number of bags to become available to customers by time t,

$F_b(t)$ = fraction of bags to arrive by time t or the probability that some arbitrarily chosen bag will arrive by time t,

$N_p F_p(t)$ = cumulative number of passengers to arrive at the baggage claim area by time t,

$F_p(t)$ = fraction of passengers to arrive by time t or the probability that some arbitrarily chosen passenger will arrive by time t.

There are two queues, a queue of bags waiting for their owners and a queue of passengers waiting for their bags. The service facility is some fictitious thing which unites a customer with his bag or bags and sends both on their way. If we neglect the time it takes a customer to remove his bag from the carousel or rack, the probability that an arbitrary bag has been removed from the carousel by time t is the probability that the bag has arrived by time t and its owner has also arrived by time t. It is reasonable to assume that the latter two events are statistically independent.

The expected cumulative number of bags to depart by time t is N_b times the probability that a bag has left, i.e.,

$$D_b(t) = \text{cumulative number of bags to leave by time } t$$
$$= N_b F_b(t) F_p(t).$$

The expected number of bags in the queue of bags is therefore

$$Q_b(t) = N_b F_b(t)[1 - F_p(t)].$$

The above relations are valid even if some passengers have more than one bag provided that the customer removes a bag from the carousel immediately even if he must still wait for a second bag. Some other aspects of this are treated in problems 3.16 and 3.17.

Problems

3.1 Draw the graphs of $D_j'(t)$ corresponding to the $D_j(t)$ of Fig. 3.3 for tandem queues with finite storage.

3.2 A traffic signal has equal red and green times, $r = g$, and a service
rate μ during green. Upstream of the signal there is a restriction
(perhaps a blocked lane) which can pass cars at a constant rate λ,
but there is room to store at most c cars between the restriction
and the signal. Determine the maximum rate at which cars can
pass through the system as a function of λ, μ, c, and g.

3.3 Cars arrive at a signalized intersection at a constant rate λ, $A(t)$
$= \lambda t$, $\lambda = 10$ cars per minute. The traffic signal is periodically
green for 30 seconds and red for 30 seconds. It starts a red at time
zero with an empty queue. During the green time, cars are
'served' (leave) at a rate 3λ. Each car travels along a road and
arrives at a second signalized intersection 30 seconds after
leaving the first. The second signal starts a red at time δ and
periodically is red for 20 seconds and green for 20 seconds. It
also serves at a rate 3λ.

Draw the arrival and departure curves with $\delta = 0$. For what
value(s) of δ is the long time average delay per car a minimum, all
other parameters remaining fixed?

3.4 A simple road network has two routes joining point A to point
B. Route 1 is an urban freeway; route 2 a city street. The
cumulative number of vehicles to arrive at point A by time t is a
given function of t, $A(t)$. The arrival rate $\lambda(t)$ rises to a single
maximum, typical of the rush hour, exceeding the service rate μ
of the freeway but less than the combined service rate of both
routes.

At capacity the travel time from A to B on route 1 is T^*; the
travel time on route 2 is $T^* + T$, independent of the flow. At
point A, drivers have the option of joining a FIFO queue for
route 1 (if there is one) or taking route 2. Each driver chooses the
route which minimizes his arrival time at B.

Construct graphs showing, as a function of t, the cumulative
number of trips to leave A on each of the routes, and the number
which reach B on each route.

3.5 A freeway begins at a city center and extends, for all practical
purposes, an infinite distance. A driver leaving the city center has
the option of using the freeway at a speed of 60 miles per hour or
various parallel city streets at a speed of 30 miles per hour. The
city streets have a capacity sufficient to carry all the traffic, if
necessary, but the freeway can carry only 4000 cars per hour.

Suppose that 5000 cars per hour wish to travel from the city
center to destinations along the freeway and that, during any

period of time, a fraction $1/[1 + (x/3)^2]$ of the drivers wish to travel a distance greater than x miles. If each driver chooses the fastest route, a queue will form on the freeway ramps. Determine how large the queue will become after a period of time sufficient for the queue length to become stable. Assume that there is sufficient room on the ramps to store whatever cars may join the queue, and that each driver has perfect knowledge of the length of the queue, the speeds, capacities, etc.

3.6 A multiple-channel server with 10 channels has a service time of one minute for each customer. Suppose there are no customers in the system at time 0 and

$$A_c(t) = \begin{cases} 5t & \text{for} \quad 0 < t < 2 \\ 10 + 15(t-2) & \text{for} \quad 2 < t < 4 \\ 40 + 5(t-4) & \text{for} \quad 4 < t. \end{cases}$$

Construct graphs of $D(t)$ and $A_s(t)$.

3.7 In a multiple-channel loss system, if a customer cannot enter the service immediately he goes away (is lost). If each server can serve only one customer at a time, then

$$D(t) = D_{qc}(t) = D_{qs}(t) = \text{cumulative number of customers}$$
$$\text{or servers to enter service,}$$

and

$$A_c(t) - D(t) = \text{number of lost customers.}$$

The relation between $A_s(t)$ and $D(t)$ is the same as illustrated in Fig. 3.8. In particular, if all service times are equal

$$A_s(t) = m + D(t - S)$$

as in (3.8). In any case, if we know $D(\tau)$ for $\tau \leq t$, we can determine $A_s(\tau)$ at least for a short time after time t, and evaluate $dA_s(t)/dt$ at time t, the rate at which servers become available. For all service times equal,

$$dA_s(t)/dt = dD(t - S)/dt.$$

The rule for entering service does not satisfy (3.9), however. We now have a rule which says that if $A_s(t) - D(t) > 0$ customers can enter service as fast as they arrive, i.e., $dD(t)/dt = dA_c(t)/dt$, but if $A_s(t) - D(t) = 0$ they cannot enter service more rapidly than new servers become available, i.e.,

$$\frac{dD(t)}{dt} = \min \left\{ \frac{dA_c(t)}{dt}, \frac{dA_s(t)}{dt} \right\}.$$

(a) Determine $D(t)$ if $A_c(t)$ is as given in problem 3.6.

(b) Show that the solution for m servers is essentially the same as in problem 2.2 with $\mu = m/s$, if $\lambda'(t)$ does not change very much in a time S.

3.8 At a ski resort, a chair lift starts operation at 9.00 a.m. All skiers are ready to ski when the lift begins operation. There are N skiers and each rides the lift to the top, skis down the hill, and is ready for another trip after a cycle time T_s. A chair on the lift goes to the top and returns empty in a time T_l. There are M equally spaced chairs on the lift which move at a constant speed, whether occupied or not. A waiting skier will always board the first available chair.

Determine the number of skiers who are waiting for the lift at time t and the number of empty seats that pass by time t, as a function of N.

3.9 Suppose that one has a finite number N of customers (machines) and m servers (repairmen) in parallel, $m < N$. A server can serve only one customer at a time in a service time of S. A time T after a customer (machine) has been served (repaired), it returns to the queue of customers (breaks) to be served (repaired) again. At time $t = 0$ all servers are available and the N customers arrive at the server at a constant rate $N/(S+T)$ for $0 < t < S + T$. Show graphically the evolution of the system and determine the long time average rate at which customers pass the service as a function of N, m, S, and T. Determine also the equilibrium fraction of time that a server is idle.

3.10 For a work conserving service serving two types of customers, let

$$s_1 A_1(t) = \begin{cases} (3/2)t & \text{for } 0 \leq t \leq 1 \\ 3/2 + (1/2)(t-1) & \text{for } 1 \leq t \end{cases}$$

$$s_2 A_2(t) = \begin{cases} 0 & \text{for } 0 \leq t \leq 1/2 \\ t - 1/2 & \text{for } 1/2 \leq t < 3/4 \\ 1/2 & \text{for } 3/4 \leq t \end{cases}$$

(there is a discontinuity in $s_2 A_2(t)$ at $t = 3/4$).

Construct graphs of $s_j D_j(t)$ under policies of (a) priority to type 1, (b) priority to type 2, (c) FIFO.

3.11 A work conserving service serves two types of customers with cumulative arrival curves $A_1(t)$ and $A_2(t)$ and service times s_1 and s_2. Show how one could graphically construct departure curves for the two types of customers using a fluid approximation, if the queue discipline is such that the number of type 1 customers in the queue $Q(t)$ is, whenever possible, equal to $1/2$ the number of customers of type 2 in the queue.

3.12 Suppose that a work conserving server serves three types of customers. If type 1 has priority over type 2 which has priority over type 3, show how, from given cumulative arrival curves $s_j A_j(t)$, one would construct the departure curves $s_j D_j(t)$. How would the curves $s_j D_j(t)$ be constructed for FIFO queue discipline?

3.13 Garbage trucks arrive at a dumping site according to a given cumulative arrival curve $A(t)$. Each truck must (1) join a queue at a weighing station to weigh the full truck, (2) dump the garbage, (3) rejoin the queue to weigh the empty truck, and (4) leave.

There is no restriction on the rate at which trucks can dump their garbage, but it takes a time T for a truck to complete the trip from the weighing station to the dump and return. The weighing station can weigh trucks (either empty or full) at a rate of μ trucks per unit time and the queue discipline is FIFO for all trucks.

Suppose that at time 0 when the weighing station opens in the morning, there is already a queue of μT full trucks and that they subsequently arrive at a constant rate of $\mu/4$. Construct graphs showing the evolution of queue lengths, waits, etc., at the weighing station.

3.14 One elevator in a bank of elevators serves floors 6, 7, 8, and 9 of a building during the morning peak when all traffic for this elevator is up-traffic from floor 1 to floors 6, 7, 8, and 9. The time to travel from floor j to $j+1$ is 2 seconds (either up or down); there is a minimum loss of 10 seconds to stop the elevators at a floor plus an additional delay of 1 second for each passenger who is either entering or leaving the elevator. Passengers arrive at floor 1 at a constant rate of λ passengers per second for each of the floors (total rate of 4λ). The elevator can carry 15 passengers per trip.

8757777ЛИ7777777

Determine the average wait per passenger (from the time he arrives until the elevator leaves floor 1) as a function of λ for each of the following strategies.

(a) Each trip is from floor 1 to 6, 7, 8, 9 and back to 1.
(b) The elevator repeats the pattern 1 to 6, 7 to 1 to 8, 9 to 1.

Also evaluate the maximum λ which can be served. Disregard the fact that passenger counts must be integer by allowing the elevator to load fractional numbers of passengers with corresponding fractional loading times.

3.15 A single airport runway is used for both take-offs and landings. The minimum time between a landing and take-off, a take-off and a landing, or two consecutive take-offs is one minute each, but the minimum time between two consecutive landings is $3/2$ minutes. The strategy for sequencing operations is to alternate take-offs and landings whenever there is a queue of both.

Requests to land at time t arrive at a rate $\lambda_L(t)$, requests to take off at a rate $\lambda_T(t)$. At $t = 0$ (6.00 a.m.), $\lambda_L(t)$ and $\lambda_T(t)$ are so low that no queue (unfulfilled requests) forms except possibly for queues of length one caused by a request arriving while the runway is occupied. Both $\lambda_L(t)$ and $\lambda_T(t)$ are increasing with time, however, as the morning rush develops.

Depending upon the relative values of $\lambda_L(t)$ and $\lambda_T(t)$, determine which queue forms first (landings or takeoffs) under the above strategy of sequencing, the values of $\lambda_L(t)$ and $\lambda_T(t)$ when the queue forms, and when the second queue forms. A queue is interpreted in the sense of a fluid approximation, disregarding queues of length one.

3.16 Suppose $F_b(t)$ and $F_p(t)$ are the distribution functions for the arrival time of bags and passengers at a carousel, a fraction p of the passengers have one bag, a fraction $1 - p$ have two bags, and the arrival time of a passenger is independent of how many bags he has. If there are N_p passengers and $N_b = N_p + (1 - p)N_p$ bags, how many customers are waiting for bags at time t? Assume that for a customer with two bags the arrival times of the two bags are statistically independent.

3.17 Two hundred bags from an airplane arrive at the baggage claim area at time $t = 0$ but they can be fed onto the conveyor belt to the carousel only at a rate of 20 bags per minute. The arrival time of a passenger at the baggage claim area is uniformly distributed over the time 0 to 10 minutes. The arrival time of a bag to the carousel and the arrival time of its owner are assumed to be statistically independent.

(a) If the carousel and the passenger claim area are sufficiently large to hold any number of bags and passengers, and passengers remove their bags from the carousel as soon as possible, how many bags are on the carousel at time t? What is the maximum number of bags on the carousel? The 'number of bags' is actually a random variable; but since the numbers are fairly large, they are to be approximated by their expectations. (b) Suppose that the carousel can hold only 32 bags. If the carousel becomes full, the conveyor belt can supply more bags as quickly as they are removed, up to a rate of 20 bags per minute, so as to maintain 32 bags on the carousel. Determine the number of bags which have arrived at the carousel and the number which have left the carousel by time t. When will all bags have been claimed?

CHAPTER 4

Stochastic models

4.1 Probability postulates

In Chapter 1 we saw that if one specified the arrival times t_j and the service times s_j for a single-channel server with FIFO queue discipline, then one could construct curves $A(t)$ and $D_q(t)$, and evaluate queue lengths, delays, etc. The purpose of any theory, however, is not to describe what did happen, but what will happen in some repetition of an experiment or in some different experiment.

It is not possible to predict exactly what will happen in some future experiment. In particular, it is virtually impossible to guarantee that the t_j and s_j will be the same as in some previous experiment. It is often true, however, that observations are at least partially reproducible. For example, some quantities such as $A(t)$ observed over corresponding time intervals of duration t on 'similar' days may be nearly equal, i.e., the differences in the observations may be small compared with the observations. This is essentially what was implied in Chapter 2 as a justification for the fluid approximations.

For other quantities, such as the s_j, individual observations may differ considerably. Yet, if one were to make n observations of some quantity X, and let $X^{(j)}$ be the jth observation, the arithmetic mean

$$(1/n) \sum_{j=1}^{n} X^{(j)}$$

may be approximately reproducible if n is sufficiently large, i.e., a set of n new observations would yield approximately the same average (indeed a set of n' observations with $n' \neq n$ may also yield approximately the same average). If this is true (or one can reasonably postulate that it should be true), then one might describe the observation in terms of a stochastic model, i.e., in terms of probabilities.

Any physical measurement of a 'deterministic' quantity, such as the mass of an electron, is approximately reproducible in a new experiment. One imagines that there is a 'true' value even though one never really determines it. One theorizes that, if measurements of arbitrary precision could be made, one could determine the true value to any arbitrary accuracy. For stochastic phenomena, one similarly imagines that there is a 'true' average which one could measure as accurately as one pleases if one could choose n arbitrarily large.

For certain physical phenomena (perhaps radioactive decay), some people believe that there are laws that are inherently stochastic. One can repeat experiments arbitrarily many times and verify the constancy of the mean, even though outcomes of single observations are unpredictable. Whether laws are inherently stochastic or only appear to be so because one does not repeat the experiment under identical conditions will probably never be resolved. To a physicist, flipping a coin is not inherently stochastic. A coin is a rigid body obeying the laws of classical dynamics. Flipping a coin under identical conditions leads to identical results. Only carelessly flipping a coin leads to its landing heads about half the time. This is a property of the person, not the coin.

In most situations in which one might wish to apply stochastic models of queues, the appropriateness of the theory is even less clear. One is not dealing with physical laws which are valid for all times; there may be people involved, and their world is always changing. One cannot repeat an experiment 'under identical conditions' as many times as one wants. Any prediction of future behavior of a system is, at best, valid only for the near future.

In addition, as with laws of physics, one can never be certain whether something is inherently stochastic or only appears so because the observer, being ignorant of certain relevant parameters, does not know how to repeat the experiment under (nearly) identical conditions. Neither does he know, therefore, if some important parameter associated with past observations will have the same value in a future observation. For example, a person may make observations of an $A(t)$ for traffic counts on many consecutive days and obtain an empirical distribution of the observed values, for a fixed t. He might even verify that this distribution is (nearly) reproducible, but then apply it to predict the future behavior for Sundays only, not realizing that an important parameter (the day of the week) had changed. Of course, in any particular application, there are certain parameters that are obviously important and others which might be important. One never has enough time or data to explore all the things that might be relevant. As a practical matter, one usually seeks to obtain models involving as few parameters as possible.

There are various types of relatively simple stochastic models of queueing phenomena that often yield reasonable predictions. There is no question, however, that any mathematical model, no matter how plausible, may fail when applied to any particular situation.

In a stochastic model, an observable which is represented by a real number, but whose value is not reproducible, is called a random variable (designated by $X, Y, \ldots$). The actual observations, designated by $X^{(j)}$, are called samples. The hypothetical quantity for which the sample mean

$$(1/n) \sum_{j=1}^{n} X^{(j)}$$

is an approximation is called the expectation, $E\{X\}$. If, in particular, the observation describes the occurrence or nonoccurrence of some event and we let $X^{(j)} = 1$ if the event occurs on the jth trial, but $X^{(j)} = 0$ otherwise, then

$$\frac{1}{n} \sum_{j=1}^{n} X^{(j)} = \text{fraction of times the event occurs}$$

and

$$E\{X\} = P\{\text{event}\}$$

is called the probability of the event.

In a typical queueing application, the basic experiment is to observe (or predict) the entire evolution of the curves $A(t)$ and $D_q(t)$ over some finite time interval (perhaps 24 hours). A single experiment, a pair of curves $A^{(j)}(t)$, $D_q^{(j)}(t)$, is called a 'realization'. If, for any quantity $X^{(j)}$ evaluated from the curves $A^{(j)}(t)$ and $D_q^{(j)}(t)$,

$$(1/n) \sum_{j=1}^{n} X^{(j)}$$

is (nearly) reproducible (it is a random variable), then $A(t)$ and $D_q(t)$ are called random functions.

For any specified times $\tau_1, \tau_2, \ldots$, the quantities $A(\tau_1), A(\tau_2), \ldots$ are random variables having a joint probability distribution which, in principle, is defined also by the equal weight to each realization of $A(t)$, i.e., for any numbers $a_1, a_2, \ldots$,

$$P\{A(\tau_1) < a_1, A(\tau_2) < a_2, \ldots\}$$
$$= \text{fraction of observations for which } A^{(j)}(\tau_1) < a_1,$$
$$A^{(j)}(\tau_2) < a_2 \ldots .$$

Other random variables such as queue lengths, waiting times, etc., are defined similarly.

If several people were to draw curves $A(t)$, $D_q(t)$ as in problem 1.1 using different sets of random digits, these could also be interpreted as separate realizations $A^{(j)}(t)$, $D_q^{(j)}(t)$ of some random functions $A(t)$, $D_q(t)$.

In most queueing models, one does not specify explicitly the probability distributions associated with the random functions $A(t)$, $D_q(t)$. To do so requires the specification of the joint probability distributions of $A(\tau_1)$, $D_q(\tau_1)$, ..., $A(\tau_m)$, $D_q(\tau_m)$ for every possible choice of $\tau_1, \tau_2, \ldots, \tau_m$ and m. This is not only mathematically difficult to describe but even more difficult to determine from observations. Instead, one usually makes some hypothesis of statistical independence among certain random variables or otherwise provides some description of properties which, in principle, defines the stochastic properties of the $A(t)$, $D_q(t)$. For example, in problem 1.1 there is a scheme by which one determines whether or not customers arrive or leave for every value of t, from which realizations of $A(t)$, $D_q(t)$ can be generated.

The properties of the $A(t)$, $D_q(t)$ are usually specified through postulated properties of the t_j, s_j, but one seeks to analyze properties such as delays or queue lengths which are more directly associated with properties of the curves $A(t)$ and $D_q(t)$. Much of the stochastic theory of queues, therefore, centers around the problem of deriving certain properties of the random functions $A(t)$, $D_q(t)$ from assumed properties of the t_j, s_j.

In most practical queueing problems one is not particularly interested in the detailed probability structure of the curves $A(t)$ and $D_q(t)$, i.e., the joint probabilities of $A(\tau_k)$, $D_q(\tau_k)$ for all τ_k, one is only interested in some crude measures of performance such as the average wait per customer, the time average queue length, or the total wait during a rush hour (or their distribution over many days). Whether or not some postulated properties of the t_j and s_j are valid is unimportant except as a means of calculating these measures.

All stochastic properties of queues, delays, etc., could, in principle, be evaluated from a collection of n realizations $A^{(j)}(t)$, $D_q^{(j)}$, (t), $j = 1, \ldots, n$, i.e., from n graphs. If one asks a sufficiently complex question – for example, what is the probability that $Q(\tau_1) < q_1$, $Q(\tau_2) < q_2$, ... and $Q(\tau_m) < q_m$? – it might take a very large number (millions) of realizations to obtain an accurate answer (because the event seldom occurs). In particular, to test the validity of any proposed model may require a very large number of realizations. Fortunately, most of the measures of performance do not change very much from one realization to the next. Although the wiggles in curves such as Fig. 2.1(a) change appreciably from one day to the next, the total area between $A^{(j)}(t)$ and $D_q^{(j)}(t)$, or the difference between this area and the deterministic approximation, over some time period typically does not change very much. If one could generate just a few curves $A^{(j)}(t)$, $D_q^{(j)}(t)$ (perhaps 10), one would likely find that the properties of interest averaged over a small sample are reproducible as accurately as is necessary for their intended use.

The most primitive question one can ask is: given a set of observations on past performance of a system during n days, what is likely to happen tomorrow if nothing is done to change the system physically? If n were sufficiently large (enormous), one would say that the $A(t)$, $D_q(t)$ curves which will obtain tomorrow are equally likely to be like any pair of curves obtained on any of the n previous days (all of which are probably different in some way), but, more important, the average wait or queue length tomorrow is equally likely to be like that observed on any of the previous days (all of which are likely to be similar).

In most cases it requires less effort to measure those things of interest (waiting times or queue lengths) on a few days and evaluate averages or distributions, than to observe arrival times, service times, etc., so as to calibrate some model from which one can calculate the distribution of queue lengths, waits, etc.

To make predictions of what would happen if the arrival curves or service rates were changed is, of course, a more difficult task. Conceptually, one must still imagine that one could generate some possible realizations $A^{(j)}(t)$, $D_q^{(j)}(t)$ for the modified system. To do this, one must have some understanding of how the properties of the $A^{(j)}(t)$, $D_q^{(j)}(t)$ relate to whatever parameters are being changed.

The most common method of doing this is to propose some detailed model of the queueing process relating the $A(t)$, $D_q(t)$ to the t_j and s_j perhaps and then make conjectures as to how the properties of the latter depend upon the modifications which are made. It may, however, be possible to speculate directly on how some new $A^{(j)}(t)$, $D_q^{(j)}(t)$ would differ from previously observed curves. If, for example, only the service is changed, one might keep the same curves $A^{(j)}(t)$ and draw some new $D_q^{(j)}(t)$ curves which one believes that a modified server would generate. If, on the other hand, next year's $A^{(j)}(t)$ will differ from last year's because of a slight anticipated growth in demand, one could simply add to each of last year's $A(t)$ curves some random cumulative curve representing the 'new' customers. Appropriate modifications in the $D_q(t)$ curves would then be drawn. In any case it is helpful to understand what various hypothetical models would predict. We will consider first some simple models for a single class of customers served by single- or multiple-channel servers.

4.2 Service and arrival distributions

From each realization $A^{(j)}(t)$, $D_q^{(j)}(t)$ one can measure many service times S_k; over many realizations one could observe a fairly large number. From these, one can define an empirical distribution function as the fraction of S_k with $S_k \leq t$, $0 \leq t < \infty$. If this distribution function is (nearly) reproducible, it can be interpreted as an approximation to some hypothetical 'true' distribution function $F_S(t)$ $= P\{S < t\}$. This will almost certainly be the case if the service times are a property of some server (rather than the customers) and the service facility does not physically change with time. If service time is a property of the customers, however, the validity would depend upon reproducibility of their collective behavior.

In most simple queueing models, it is postulated that whenever a new service time is selected in the process of generating some realization $D_q(t)$, it is selected at random from the distribution $F_S(t)$ independent of any past history of events. Equivalently, a new S_j is equally likely to have a value (nearly) equal to any previously observed S_k. The joint distribution for any sequence of service times is postulated to have the form

$$P\{S_1 \leq t_1, S_2 \leq t_2, \ldots\} = F_S(t_1)F_S(t_2)\ldots,$$

i.e., they are statistically independent.

There are all sorts of conditions under which this hypothesis would be clearly false, but unless one has strong reasons for believing that statistical dependencies do exist, one will usually assume that they do not. To assume otherwise would necessitate experimental measurements to determine the nature of the dependence, and also complicate any theory.

There are obviously statistical dependencies among service times for non-work-conserving systems in which the service time of one customer may depend upon the type (or service time) of a preceding or following customer. In multiple-channel servers, there may also be various types of interactions among individual servers.

STOCHASTIC MODELS   111

A more subtle source of dependence may arise, however, from the fact that differences in service times may be attributed to the customer rather than the server. The customer may know approximately how much service time he needs, so his service time is not really 'stochastic', not with the distribution function $F_S(t)$ at least. If one wished to know his service time, one could ask him. Just because the observer is ignorant of this does not mean that the system does not know.

If customers of known service times are selected in random order from a large population, the service times will appear as if each were selected from a common distribution $F_S(t)$. If, however, the same customer appears every day at nearly the same time, this may affect the total delay but have little effect upon certain statistical tests for independence of successive S_k. Just because a single realization of $D_q(t)$ behaves as if it were generated from statistically independent service times does not mean that repetition of the experiment will yield a new realization with properties like those that would result from selecting *new* service times at random from the same service distribution.

The stochastic properties of arrival processes encountered in typical applications are usually much more difficult to model than the properties of the service. For one thing, the arrival rate $\lambda(t) = dE\{A(t)\}/dt$, is usually time dependent whereas the properties of the server are usually stationary. Furthermore, the probability structure of the process is likely to be quite complex. Arrivals at one server may have passed through other servers (which cause inter-actions between arrival times of successive customers). There may exist a known or effective schedule of arrivals. For example, aircraft arriving at an airport or buses at a bus stop may appear to arrive with statistically independent times between successive arrivals; yet the number of arrivals over a long time (24 hours, say) may be known exactly. Even for highway traffic there may be an effective schedule in that the same cars appear every day at nearly the same time. On the other hand, there are all sorts of events large and small occurring on any day which can cause more customers to arrive on one day than another. One can observe it, but one cannot explain why.

There are situations, however, in which customers arrive at a service facility via different paths and have had no chance to interact with each other prior to their time of arrival. If we number all potential customers according to some arbitrary scheme (unrelated to their actual arrival times), then it may be reasonable to assume that the arrival time of the customer numbered j is statistically independent of the arrival time of any other customer. Not all customer arrival times are necessarily sampled from the same distribution, however.

Suppose, for example, that an airline knows that there are N passengers who are expected to arrive at an airport to check in for some scheduled flight. Although certain passengers always arrive early and others always arrive at the last minute and the passengers know their own habits, the airline does not know the habits of the customers who have made reservations. From a very large sample of passengers for previous flights, however, the airline has determined an (empirical) distribution function of arrival times of passengers relative to the scheduled flight time, i.e.,

$$F_A(t) = P\{\text{arrival time of a customer} \leq t\}.$$

This flight is not a commuter flight which has regular customers whose habits could, in principle, have been predicted from past observations. We assume that the N passengers expected for this flight are, in effect, sampled from a very large population of potential customers having an arrival distribution $F_A(t)$. Suppose also that all passengers are traveling alone; they did not come to the airport in the same car or bus. If we number the N customers in some arbitrary order (alphabetically by name) we might further assume that

$$P\{\text{customer 1 arrives before time } t_1 \text{ and customer 2} \atop \text{arrives before time } t_2, \text{ and } \ldots\} = F_A(t_1)F_A(t_2)\ldots \qquad (4.1)$$

i.e., the arrival times of customers $1, 2, \ldots$ are statistically independent and identically distributed.

The original numbering of passengers is irrelevant to the queueing process. We do not care who arrived by some time but only how many arrived. In particular, what is the probability that k passengers arrived by time t, i.e., $A(t) = k$? The evaluation of this is one of the classic exercises in elementary probability theory. If we think of a customer arriving by time t as success and after time t as failure, with probabilities $F_A(t)$ and $1 - F_A(t)$, respectively, then we are asking: what is the probability of k successes in N trials? Thus

$$P\{A(t) = k\} = \frac{N!}{(N-k)!k!}[F_A(t)]^k[1 - F_A(t)]^{N-k}, \qquad (4.2)$$

a binomial distribution with 'parameter' $F_A(t)$.

This describes the (marginal) probability distribution of the random variable $A(t)$ for each t (it does not describe the joint probability distribution for $A(\tau_1), A(\tau_2), \ldots$). From (4.2) one finds

$$E\{A(t)\} = \sum_{k=0}^{N} kP\{A(t) = k\} = NF_A(t) \qquad (4.3)$$

$$\mathrm{Var}\{A(t)\} = \sum_{k=0}^{N} [k - NF_A(t)]^2 P\{A(t) = k\}$$

$$= NF_A(t)[1 - F_A(t)] \equiv \sigma^2_{A(t)}. \qquad (4.4)$$

The expected curve, $E\{A(t)\}$, is nothing more than the multiple N of the distribution function for a single passenger. The standard deviation $\sigma_{A(t)}$ is a measure of the fluctuation in the value of $A(t)$ from one realization to the next and is proportional to $N^{1/2}$. It vanishes for $F_A(t) \to 0$ (no passengers have arrived) and also for $F_A(t) \to 1$ (all passengers have arrived).

From a known distribution $F_A(t)$ one can draw, as in Fig. 4.1, curves of $E\{A(t)\}$ (solid curve) and $E\{A(t)\} \pm \sigma_{A(t)}$ (broken lines). The interpretation of the latter curves is that we would expect most realizations of $A(t)$ to stay between these curves most of the time. Since $E\{A(t)\}$ is proportional to N but $\sigma_{A(t)}$ is proportional only to $N^{1/2}$, $\sigma_{A(t)}$ will be small compared with $E\{A(t)\}$ for $N \gg 1$.

If we rescale Fig. 4.1 by N and consider realizations of $A(t)/N$, this, in effect, is the empirical distribution function for the N observations of arrival times. There is a very extensive literature on the stochastic properties of empirical distribution functions as estimators of the true

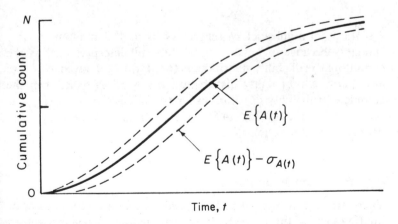

Figure 4.1 *Binomial distribution of arrivals*

distribution function $F_A(t)$. This is a standard topic in almost any textbook on mathematical statistics.

Equation (4.2) does not completely describe the arrival process; it is a consequence of (4.1) but not conversely. A more complete description of the arrival curve $A(t)$ is as follows. For any set of times $\tau_0 < \tau_1 < \tau_2 < \ldots \tau_m$ the probability that a jth passenger arrives between times τ_k and τ_{k+1} is $F_A(\tau_{k+1}) - F_A(\tau_k)$, $k = 0, 1, \ldots, m-1$; the probability that he arrives before time τ_0 or after τ_m is $F_A(\tau_0)$ or $1 - F_A(\tau_m)$, respectively. The probability that k_0 passengers arrive before time τ_0, k_1 arrive between times τ_0 and $\tau_1, \ldots$, and k_{m+1} arrive after time τ_m is given by the multinomial distribution

$$[F_A(\tau_0)]^{k_0} [F_A(\tau_1) - F_A(\tau_0)]^{k_1} \ldots [F_A(\tau_m) - F_A(\tau_{m-1})]^{k_m}$$
$$\times [1 - F_A(\tau_m)]^{k_{m+1}} (k_0 + k_1 + \ldots + k_{m+1})!/(k_0! \, k_1! \ldots k_m! \, k_{m+1}!)$$
(4.5)

provided that

$$k_0 + k_1 + \ldots k_{m+1} = N. \qquad (4.5a)$$

There is zero probability if the k_j do not satisfy (4.5a).

This describes completely the stochastic properties of the arrival process, disregarding the identity (i.e., the numbering) of the N customers. Expression (4.5) is valid for arbitrary choices of m, τ_0, $\tau_1, \ldots, \tau_m$ and describes the joint probability distribution of $A(\tau_0)$, $A(\tau_1), \ldots, A(\tau_m)$ since it determines

$$P\{A(\tau_0) = k_0, A(\tau_1) = k_0 + k_1, \ldots, A(\tau_m) = k_0 + k_1 + \ldots k_m\}$$
$$\text{for any } k_0, k_1, \ldots, k_m \geq 0.$$

If the total number of passengers, N, is itself a random variable (possibly because of no-shows), we can still interpret (4.5) as the conditional probability distribution for $A(\tau_0)$, ..., given N. If we also know a probability distribution for N, $P\{N = n\}$, then the complete probability distribution for the values $k_0, k_2, \ldots k_{m+1}$ is obtained by multiplying (4.5) by $P\{N = k_0 + \ldots + k_{m+1}\}$ and dropping the restriction (4.5a).

4.3   A Poisson process

There are several types of arrival processes that are special cases or limiting cases of the above. Suppose that, instead of having a known finite number of customers who are expected to arrive within some finite time before the departure of an aircraft, we have cars arriving at an intersection or passengers arriving for buses which leave every few minutes. We have a source of N potential customers, N being very large, perhaps all the customers who might arrive anytime during the day, but during any small time interval τ_{k-1} to τ_k (a few minutes or less) there is a very small probability that any particular customer will arrive, i.e.,

$$F_A(\tau_k) - F_A(\tau_{k-1}) \ll 1.$$

If we start our observations at time τ_0 and stop at time τ_m, we are really interested only in the joint distribution of counts within the time intervals $\tau_k - \tau_{k-1}$ not with $A(\tau_0)$ or $N - A(\tau_m)$, the number of arrivals before or after the period of observation; but we assume that the number observed during the time τ_0 to τ_m is (almost certainly) small compared with N.

To eliminate the distribution of $A(\tau_0)$ from (4.5) we can either sum (4.5) over all k_0 and k_{m+1} satisfying (4.5a) to obtain the joint (marginal) distribution of the $A(\tau_{k+1}) - A(\tau_k)$, $k = 0, 1, \ldots, m-1$ averaged over all ways in which the counts outside the period of observations are partitioned between times before τ_0 or after time τ_m; or we could evaluate the conditional distribution of the $A(\tau_k) - A(\tau_{k-1})$ for given $A(\tau_0) = k_0$ and N. In either case, we are interested in any limiting behavior of this distribution when we let $N \to \infty$ and $F_A(\tau_{k+1}) - F_A(\tau_k) \to 0$ in such a way that the expected number of arrivals in the kth interval

$$E\{A(\tau_{k+1}) - A(\tau_k)\} = N[F_A(\tau_{k+1}) - F_A(\tau_k)] \to \beta_{k+1}$$

has a finite limit β_{k+1}.

One can show that there is such a limit distribution, namely

$$P\{A(\tau_1) - A(\tau_0) = k_1, A(\tau_2) - A(\tau_1) = k_2,$$
$$\ldots, A(\tau_m) - A(\tau_{m-1}) = k_m\}$$
$$= \frac{\beta_1^{k_1} e^{-\beta_1}}{k_1!} \times \frac{\beta_2^{k_1} e^{\beta_2}}{k_2!} \times \ldots \times \frac{\beta_m e^{-\beta_m}}{k_m!} \qquad (4.6)$$

which is independent of N and depends only upon the expected number of arrivals β_k within each time period. Equivalently, the probability of k_l arrivals in the time interval (τ_{l-1}, τ_l) has a Poisson distribution

$$P\{A(\tau_l) - A(\tau_{l-1}) = k_l\} = \frac{\beta_l^{k_l} e^{-\beta_l}}{k_l!} \qquad (4.6a)$$

and the number of arrivals in this time interval is statistically independent of the number in any other non-overlapping time interval. Since it is also independent of the count at time τ_0, we could start counting from time τ_0 with $A(\tau_0) = 0$.

A mathematically simpler way to derive (4.6), which avoids taking such limits, is to postulate that N, instead of having a given large value N_0, is a random variable itself having a Poisson distribution with mean N_0:

$$P\{N = n\} = N_0^n e^{-N_0}/n!$$

i.e.,

$$P\{N = k_0 + k_1 + \ldots + k_{m+1}\} = \frac{N_0^{k_0 + k_1 + \ldots + k_{m+1}} e^{-N_0}}{(k_0 + k_1 + \ldots + k_{m+1})!}.$$

With this distribution for N, one can see immediately from (4.5) that all the random variables $A(\tau_0)$, $A(\tau_{k+1}) - A(\tau_k)$ and $N - A(\tau_m)$ will be statistically independent, and each will have (exactly) a Poisson distribution with the appropriate mean value.

For $N_0 \gg 1$, the Poisson distribution of N is relatively narrow $(\sigma_N/N_0 = N_0^{-1/2})$. It clearly should make little difference to the counts in τ_0 to τ_m whether we assume that N is fixed or has a narrow distribution. In fact, the appropriate value of N_0 is usually not known in practical applications, and neither is $F_A(\tau_{k+1}) - F_A(\tau_k)$. The parameters in the Poisson distribution $E\{A(\tau_{k+1}) - A(\tau_k)\}$ are measured (or predicted) directly.

One can show either directly from (4.6a) or by taking an appropriate limit of (4.4), that for a Poisson distribution of $A(t)$

$$\text{Var}\{A(t)\} = E\{A(t)\}.$$

Since $E\{A(t)\}$ is monotone increasing with t, so is $\sigma_{A(t)}$. The analogue of Fig. 4.1 for a Poisson distribution of $A(t)$ is shown in Fig. 4.2.

An arrival process $A(t)$ which satisfies (4.6) for all values of $\tau_0, \tau_1, \ldots, \tau_m$ and m is called a Poisson process or more commonly an inhomogeneous Poisson process. In many books on stochastic processes, the term Poisson process is used in a more restrictive sense to mean a homogeneous Poisson process. A homogeneous Poisson process has the further property that $E\{A(t)\}$ is a linear function of t: i.e., the arrival rate

$$\lambda(t) = dE\{A(t)\}/dt = \lambda$$

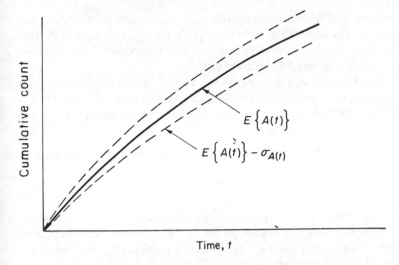

Figure 4.2   *Poisson distribution of arrivals*

is constant. Thus in (4.6)

$$\beta_l = E\{A(\tau_l) - A(\tau_{l-1})\} = \int_{\tau_{l-1}}^{\tau_l} \lambda(t)\mathrm{d}t = (\tau_l - \tau_{l-1})\lambda. \quad (4.7)$$

A homogeneous Poisson process has the important property that

$$P\{\text{no arrivals in time } \tau\} = \mathrm{e}^{-\lambda\tau}, \quad (4.8)$$

independent of how many arrivals may occur in any time intervals outside the one in question. In particular, if time 0 is an arrival time

$$P\{\text{no arrivals between time 0 and } \tau \text{ if there is one at 0}\}$$
$$= P\{\text{time until the next arrival} > \tau\} = \mathrm{e}^{-\lambda\tau}.$$

If we let T denote the random time between successive arrivals, then

$$F_T(\tau) = P\{T \le \tau\} = 1 - \mathrm{e}^{-\lambda\tau}; \quad (4.9)$$

T is exponentially distributed with parameter λ independent of the previous arrival times.

It is quite common in the analysis of queues to suppose that the arrival process is a homogeneous Poisson process, partly because it is approximately correct for some real systems but mostly because it is mathematically convenient. In addition to the fact that it is generally more difficult to deal with time-dependent than with time-independent phenomena, for the homogeneous Poisson process one can usually exploit the facts that (a) the numbers of arrivals in non-overlapping time intervals are statistically independent and (b) so are the time intervals between arrivals.

For the inhomogeneous Poisson process (a) is still true, but (b) is not. If, for example, arrivals occur at times 0, T_1 and T_2, the distribution of the time between the first and second arrivals, given T_1, is

$$P\{T_2 - T_1 \le \tau | T_1 = t_1\} = 1 - P\{\text{no arrivals between } t_1 \text{ and } t_1 + \tau\}$$

$$= 1 - \exp\left(-\int_{t_1}^{t_1 + \tau} \lambda(t)\mathrm{d}t\right).$$

Except for $\lambda(t) = \lambda$, this distribution will depend upon the time t_1.

One can formally transform an inhomogeneous Poisson process into a homogeneous Poisson process by imagining that one has a clock which measures a new 'time'

$$t'(t) = \int_0^t \lambda(\tau)\mathrm{d}\tau.$$

Relative to this clock, the expected number of arrivals in 'time' t' is t', and the arrival process is a homogeneous Poisson process of 'rate' 1. Unfortunately this trick of distorting the time axis is not very helpful in the analysis of queues because it complicates the process of departures.

For the more general process described by (4.5) with a distribution for N other than a Poisson distribution, neither (a) nor (b) is true. If N is known, for example, then the number of arrivals before time τ and the number of arrivals after time τ are obviously statistically dependent (their sum must be N). The joint probability distribution of the times between arrivals is quite complicated.

4.4 Robustness of the Poisson distribution

The most important and surprising theorems in probability theory are the ones which deal with the limit distributions of certain random variables which are functions of a large (infinite) number of other random variables. If, for example, $X^{(j)}$ are independent and identically distributed random variables, then

$$\overline{X} = (1/n) \sum_{j=1}^{n} X^{(j)}$$

has a singular limiting distribution $\overline{X} \to E\{X\}$ for $n \to \infty$ with probability one (law of large numbers), whereas

$$n^{1/2}[\overline{X} - E\{X\}]/\sigma_x$$

has a normal distribution with mean 0, variance 1 for $n \to \infty$ (central limit theorem). The surprising fact is that these limit distributions do not depend upon the distribution of X, only on its first and second moments (assumed to be finite). Even if the $X^{(j)}$ are not statistically independent or identically distributed, but satisfy other weaker conditions, these limit theorems may still be true. There are other limit distributions associated with $\max_j X^{(j)}$ (extreme value distributions) which are also insensitive to the distribution of the $X^{(j)}$.

The Poisson distribution for the counts of events in some time interval is another example of a limit distribution whose properties are rather insensitive to the detailed mechanisms which cause the events. This was illustrated in part in the last section by the fact that there was a limiting distribution for $N \to \infty$. The Poisson distribution arises, however, under much more general conditions.

Although it is impossible in any practical application to test whether or not any mechanisms which generates an arrival process statisfy certain properties which would lead to a Poisson process or even to test directly whether the arrival process itself satisfies all the properties of a Poisson process, it is helpful to understand which types of mechanisms generate Poisson processes or processes that may be similar to a Poisson process. If, then, one has reason to believe that the necessary postulates might be true, one might be content to apply only a few simple tests to confirm one's expectations.

There were essentially two types of postulates used in Sections 4.2 and 4.3 leading to the Poisson distribution of counts. One was the statistical independence of the arrival times of different customers; the other was that each customer had a distribution of arrival time corresponding to that of a customer selected at random from a very large population of potential customers. All customers, in effect, had the same arrival time distribution, but it was highly unlikely that any particular customer would arrive during any period of observation. It is obvious that various forms of statistical dependencies can cause deviations from a Poisson arrival process so we do not expect to relax this assumption very much. The second assumption, however, is often quite unrealistic and, to some extent, unnecessary.

For many arrival processes (passengers arriving for buses, cars passing some point on a highway, customers arriving at an office, bank, or whatever) a significant fraction of the customers appear every day (or at least have a nonzero probability of appearing on any given day) and they appear at approximately the same time every day (perhaps within 10 minutes or so). To model this, suppose we assume that each customer has his own distribution of arrival time, i.e., if T_j^* is the arrival time of customer j,

$$F_j(\tau) = P\{T_j^* < \tau\}.$$

If a new day is considered to start at midnight, say, then for $\tau = 24$ hours, $F_j(\tau)$ need not be 1 (some customers do not appear that day) but neither is it arbitrarily small, at least for some fraction of the population of potential customers. We also assume that the T_j^* are statistically independent, i.e., one customer's T_j^* sampled from his distribution is unrelated to the T_k^* sampled from the kth distribution.

We consider first the probability $P_0(\tau, t)$ that no customer arrives during a time t to $t + \tau$. Since the T_j^* are statistically independent,

$$P_0(\tau, t) = P\left\{\begin{matrix} \text{customer 1 does not arrive,} \\ \text{customer 2 does not arrive} \ldots \end{matrix}\right\}$$

$$= \prod_{j=1}^{N} \{1 - [F_j(t+\tau) - F_j(t)]\}. \tag{4.10}$$

Suppose now that for each j, the distribution $F_j(t)$ is spread over such a range that even for a customer who is likely to arrive sometime near time t, the uncertainty of his arrival time is large compared with τ (if the standard deviation of T_j^* is 10 min, say, then τ is one minute or less). Specifically, we assume that

$$\varepsilon_j = F_j(t+\tau) - F_j(t) \ll 1 \quad \text{for all } j \tag{4.11}$$

but N is sufficiently large that the expected number of arrivals in $(t, t+\tau)$

$$\sum_{j=1}^{N} \varepsilon_j = \sum_{j=1}^{N} [F_j(t+\tau) - F_j(t)] = E\{A(t+\tau) - A(t)\} = \beta \tag{4.12}$$

is not small compared with 1.

Each factor of (4.10) can be written as

$$1 - \varepsilon_j = \exp[\ln(1 - \varepsilon_j)] = \exp(-\varepsilon_j - \varepsilon_j^2/2 \ldots).$$

Therefore

$$P_0(\tau, t) = \exp\left[-\sum_{j=1}^{N} \varepsilon_j - \sum_{j=1}^{N} \varepsilon_j^2/2 \ldots\right] \simeq e^{-\beta}.$$

Since $\varepsilon_j \ll 1$, the term

$$\sum_{j=1}^{N} \varepsilon_j^2/2$$

is certainly small compared with β. If it is not also small compared with 1, then $e^{-\beta}$ is negligibly small anyway.

The probability that customer j arrives in $(t, t+\tau)$ but no others is

$$\varepsilon_j \prod_{k \neq j} (1 - \varepsilon_k) = \frac{\varepsilon_j}{(1 - \varepsilon_j)} P_0(\tau, t).$$

Therefore

$$P_1(\tau, t) = \sum_{j=1}^{N} \left(\frac{\varepsilon_j}{1 - \varepsilon_j} \right) P_0(\tau, t) \simeq P_0(\tau, t) \sum_{j=1}^{N} \varepsilon_j = \beta e^{-\beta}.$$

Similarly, the probability that customers j and k arrive in $(t, t+\tau)$ but no others is

$$\frac{\varepsilon_j \varepsilon_k}{(1 - \varepsilon_j)(1 - \varepsilon_k)} P_0(\tau, t)$$

and so

$$P_2(\tau, t) = \sum_{1 < j < k < N} \frac{\varepsilon_j \varepsilon_k}{(1 - \varepsilon_j)(1 - \varepsilon_k)} P_0(\tau, t)$$

$$\simeq \frac{1}{2} P_0(\tau, t) \left\{ \left[\sum_{j=1}^{N} \varepsilon_j \right]^2 - \left[\sum_{j=1}^{N} \varepsilon_j^2 \right] \right\}$$

$$\simeq \frac{\beta^2}{2} e^{-\beta}.$$

The obvious continuation of this procedure gives the Poisson distribution

$$P_k(\tau, t) = \frac{\beta^k}{k!} e^{-\beta}.$$

Under the same assumptions, one can also show that the number of arrivals in non-overlapping subdivisions of the time $(t, t+\tau)$ are statistically independent. Thus, the arrival process appears 'locally' like a Poisson process. Also if β is proportional to τ, even though the individual ε_j are not, the process behaves like a homogeneous Poisson process with independent and exponentially distributed times between arrivals. It does not follow, of course, that the same is true over a time interval larger than τ. Since ε_j is an increasing function of τ, (4.11) is likely to be false if τ is too large.

4.5 Deviations from a Poisson process

In the analysis of queues one may be interested in the arrival and departure processes over a wide range of possible time scales. If $\lambda(t)/\mu$ is considerably less than 1, it is unlikely that queues will be very large (they are usually 0 or 1) and any analysis would obviously focus on the behavior of individual customers on a time scale comparable with the interarrival or service times. If, however, $\lambda(t)/\mu$ is close to or greater than one, queues will persist over a time large compared with the service time and the size of the (stochastic) queue will be determined mostly by fluctuations in the cumulative counts involving a large number of arrivals.

Most processes encountered in applications, which have properties similar to Poisson processes, also have various types of distortions. They do not quite satisfy the appropriate postulates and one must make certain corrections depending upon the scale of time which is relevant.

For $\lambda(t)/\mu$ small ($\lesssim 1/2$ say), the probability of having a nonzero queue is essentially determined by the probability that, after one customer has entered service, a second (or third) customer arrives before the service of the first is completed. Once a queue has formed, it is likely to last only a few service times at most. Since this behavior is quite sensitive to properties of the interarrival times, one must look for possible deviations from an exponential distribution and statistical dependencies between neighboring interarrival times.

Possible interactions between individual customers are the most likely sources of errors. If customers passed through another server *en route*, the times between successive arrival of customers who had been in a queue of the input server would be determined by the service times of the input server. For example, the 'expected times of arrival' of aircraft at an airport may seem to have an exponential distribution of interarrival times but, as the aircraft enter the final approach path, the traffic controller imposes a minimum time (at least 90 seconds) between landings. The traffic controller acts like an input server with a service time of at least 90 seconds. Cars or buses traveling along a highway must also maintain some minimum time headway to avoid collision.

For some other arrivals processes one might find an excess of short headways as compared with a Poisson process. For example, some passengers arriving at a bus stop may be traveling together. A reasonable generalization of the Poisson process might be one in which batches of customers arrive like a Poisson process but each batch contains a random number of (nearly) simultaneous arrivals (compound Poisson process). On a highway, even though there is a deficiency of short headways (less than one second) because cars cannot be too close together, there is usually an excess of headways less than about three seconds because a car wishing to pass another may be delayed. The cars travel like a pair for a while (but not too close).

Most practical queueing problems deal with queues of moderate to large size (5 to 1000) simply because any efficiently designed facility should be congested at some times. The types of deviations from a Poisson process which are important in such cases are quite different from those for light traffic. Whether or not individual headways are exponentially distributed is not important in itself; it makes little difference to the value of a queue if a single customer arrives a little earlier or later than expected. The important properties are variation in the counts of arrivals or services over moderately long times as measured perhaps by $\mathrm{Var}\{A(t)\}$ and $\mathrm{Var}\{D(t)\}$.

For a Poisson process $\mathrm{Var}\{A(t)\} = E\{A(t)\}$, but for many processes encountered in applications $\mathrm{Var}\{A(t)\}/E\{A(t)\}$ may differ considerably from 1 even though interarrival times may seem to be exponentially distributed and successive headways statistically independent. There are several common types of effects that can cause this.

In the situation described in Section 4.4, the time τ may be so large that $\varepsilon_j \simeq 1$ for some j. Aircraft arriving at an airport, for example, have a schedule. On a scale of 5 minutes or less, the arrivals may seem to describe a Poisson process because the probability of any particular aircraft arriving within a 5-minute period is small even within 5 minutes of the scheduled time (i.e., $\varepsilon_j \ll 1$), but it is likely to arrive within 30 minutes of its scheduled time ($\varepsilon_j \simeq 1$). For such a process $\mathrm{Var}\{A(t)\}/E\{A(t)\}$ is typically less than 1. If the T_j^* are statistically independent,

$$E\{A(t+\tau) - A(t)\} = \sum_{j=1}^{N} \varepsilon_j \qquad (4.13)$$

and

$$\mathrm{Var}\{A(t+\tau) - A(t)\} = \sum_{j=1}^{N} \varepsilon_j(1 - \varepsilon_j) \leq \sum_{j=1}^{N} \varepsilon_j. \qquad (4.14)$$

In particular, any ε_j which is (nearly) 1 contributes to the mean but not to the variance.

Suppose, on the other hand, that customers arrive in batches of size B_k; the batches form a Poisson process, and the B_k are statistically independent and identically distributed. If $A(t)$ represents the number of arrivals and $A'(t)$ the number of batches, then

$$A(t) = \sum_{k=1}^{A'(t)} B_k,$$

$$E\{A(t)\} = E\{A'(t)\}E\{B\}, \qquad (4.15)$$

but

$$\mathrm{Var}\{A(t)\} = E\{A'(t)\}\,\mathrm{Var}\{B\} + E^2\{B\}\,\mathrm{Var}\{A'(t)\}.$$

Since, for a Poisson distribution of $A'(t)$, $\mathrm{Var}\{A'(t)\} = E\{A'(t)\}$

$$\frac{\mathrm{Var}\{A(t)\}}{E\{A(t)\}} = \frac{E\{B^2\}}{E\{B\}} = E\{B\}[1+C^2(B)] \geq 1. \qquad (4.16)$$

If, for example, all batches were the same size, $B_j = B$, then $A(t) = BA'(t)$ would have a Poisson distribution on multiples of B and $\mathrm{Var}\{A(t)\} = B^2\,\mathrm{Var}\{A'(t)\}$, $E\{A(t)\} = BE\{A'(t)\}$, $\mathrm{Var}\{A(t)\}/E\{A(t)\} = B$.

In many cases, a process might look like a Poisson process on any particular day (even a homogeneous process over some finite time interval) but for unknown or unpredictable reasons (possibly the weather) the 'demand' is somewhat different on different days. Over many days the process might behave as if it were a homogeneous Poisson process with a rate Λ that varies randomly from day to day (but is constant on the same day). For such a process

$$E\{A(t)\} = E\{\Lambda t\} = tE\{\Lambda\} \qquad (4.17)$$

$$\mathrm{Var}\{A(t)\} = E\{\mathrm{Var}\{A(t)|\Lambda\}\} + \mathrm{Var}\{E\{A(t)|\Lambda\}\}$$
$$= E\{\Lambda t\} + \mathrm{Var}\{\Lambda t\}$$

and

$$\frac{\mathrm{Var}\{A(t)\}}{E\{A(t)\}} = 1 + \frac{\mathrm{Var}\{\Lambda t\}}{E\{\Lambda t\}} = 1 + tE\{\Lambda\}C^2(\Lambda). \qquad (4.18)$$

Chances are that Λ does not change very much from one day to the next (among similar days) in the sense that $C^2(\Lambda) \ll 1$. This means that on a time scale involving only a few arrivals, $tE\{\Lambda\}$ comparable with 1, $\mathrm{Var}\{A(t)\}/E\{A(t)\}$ is close to 1 and furthermore headways may appear to be exponentially distributed even if one considers headways on different days. Any test for statistical independence of neighboring headways is likely also to pass. But this variance to mean ratio increases linearly with t. For a time involving many arrivals $t E\{\Lambda\} \gg 1$ and, in particular, for $t E\{\Lambda\} C^2(\Lambda)$ comparable with 1 or more, the variance to mean ratio will differ appreciably from 1. If queues persist for a time of this magnitude, the fluctuations from day to day will originate more from the variation in the Λ than from variations in the headways.

4.6 The normal approximation

One of the important properties of nearly all the processes described above is that, during any time interval for which the expected number of arrivals is large compared with 1, the distribution for the number of arrivals is approximately normal. That this is true for particular types of processes is usually described in the probability literature under the general heading of 'central limit theorems'. To prove such theorems, one considers some suitable sequence of random variables Y_n, all of which have some special type of distribution (binomial, for example). The theorems then state that the distribution functions of the Y_n approach a normal distribution for $n \to \infty$.

In real applications, however, one seldom deals with any sequence of random variables (certainly not an infinite sequence). There is usually only one random variable, but we assume that it is sufficiently far out in some hypothetical sequence that its distribution is close to the limit distribution ($n \to \infty$). The final test of suitability, however, is whether or not it yields results of the desired accuracy.

If Y is normal with

$$E\{Y\} = m \quad \text{and} \quad \text{Var}\{Y\} = E\{(Y-m)^2\} = \sigma^2,$$

it has a probability density

$$P\{y < Y < y + \mathrm{d}y\} = \frac{\mathrm{d}y}{\sqrt{2\pi}\sigma} \exp\left[-\frac{(y-m)^2}{2\sigma^2}\right].$$

and a distribution function

$$F_Y(y) = P\{Y < y\} = \int_{-\infty}^{y} \frac{\exp\left[-\dfrac{(y'-m)^2}{2\sigma^2}\right]}{\sqrt{2\pi}\sigma} \, \mathrm{d}y'. \quad (4.19)$$

We are mostly concerned here with integer valued random variables which count customers. An integer valued random variable Y cannot have a probability density, but it does have a distribution function.

Even though the exact distribution function increases with y by jumps at integer y, the jumps may be so small that the step function distribution function stays everywhere close to the smooth differentiable normal distribution function.

It is in this sense that the number of arrivals during some sufficiently long time is approximately normal. Whereas in the previous sections we treated customers as if they formed a fluid with a non-random, deterministic arrival and/or departure rate, we now recognize that there is an intermediate level of approximation between this deterministic fluid approximation and an exact stochastic treatment. In this intermediate approximation we can still think of the customer count as an amount of fluid, disregarding the discrete nature of the counts; we treat certain random variables as if they were normal and had a continuum of possible values, but the continuous fluid is subject to random fluctuations.

For most processes that one encounters in applications, $A(t)$ will be approximately normal if $E\{A(t)\} \gg 1$, and $\mathrm{Var}\{A(t)\}/E\{A(t)\}$ will be comparable with 1 (within a factor of 2 or 3) over a wide range of values for $E\{A(t)\}$. For $E\{A(t)\} \gg 1$, it follows that

$$E\{A(t)\} \gg \sigma_{A(t)} \gg 1. \qquad (4.20)$$

Thus, at any time one could justify using a fluid approximation neglecting $\sigma_{A(t)}$ compared with $E\{A(t)\}$, the use of a normal distribution represents a second approximation in which the discreteness of counts, 1, is neglected compared with $\sigma_{A(t)}$.

These normal approximations for the marginal distribution of $A(t)$ can also be generalized to describe approximate joint distributions of $A(\tau_0)$, $A(\tau_1) - A(\tau_0)$, ... If $E\{A(\tau_k) - A(\tau_{k-1})\} \gg 1$, the $A(\tau_k) - A(\tau_{k-1})$ will each have an approximate normal marginal distribution and together they will have a joint multidimensional normal distribution. For the Poisson process, these random variables will be statistically independent and the joint normal distribution will be simply the product of normal marginal distributions. It is, of course, the differences $A(\tau_k) - A(\tau_{k-1})$ that are statistically independent, not the $A(\tau_k)$ themselves. The $A(\tau_k)$ will also have a joint normal distribution but with nonzero covariances.

In the probability literature, a process $A(t)$ having joint normal probability distributions for $A(\tau_k)$ is called a normal or Gaussian process. A process for which the $A(\tau_k) - A(\tau_{k-1})$ are statistically independent is called a process of independent increments. One with independent and normally distributed increments is called a

Brownian motion, a diffusion process, or a Wiener process. The literature on such subjects is very extensive.

For the multinomial distributions of Section 4.2, there will be a nonzero covariance between the random variables $A(\tau_k) - A(\tau_{k-1})$ because of the fixed value of N, but the joint normal distribution for these will still be quite simple. One can obtain the joint distribution by first imagining that N has a Poisson distribution so that all component counts are statistically independent Poisson (or independent normal), and then divide this distribution by the probability that $N = N_0$ to obtain the conditional distribution for $N = N_0$.

For more general types of arrival processes one may have nontrivial covariances between the counts, but in all cases the complete joint normal distribution is uniquely defined by the covariances and expectations, all of which can usually be estimated directly from observation of realizations of the $A(t)$.

4.7 The departure process

The usual assumption made regarding the departure process is that it is generated from service times which are independent random variables with a distribution function $F_S(t)$. Even though the arrival and departure processes may involve independence postulates, the two processes are (except in very special cases) stochastically quite different. For the arrival process described in Sections 4.2 and 4.3, it is the arrival times of different customers that are independent; but for the departure process from a single-channel server working without interruption (when the queue is nonzero), it is the time intervals between successive departures that are independent. Such a process is called a renewal process in the probability literature.

For a single-channel server, suppose that at time zero we have a large queue that will remain positive for many service times and that a new service is about to start. The first customer will leave the service at time S_1, the second at time $S_1 + S_2$, etc. The curve $D_q(t)$ is a step function with unit steps at these times as illustrated in Fig. 4.3.

Whereas, for certain types of arrival processes $A(t)$, the vertical increments $A(\tau_k) - A(\tau_{k-1})$ are independent random variables (possibly with a Poisson distribution), for $D_q(t)$ it is the horizontal increments that are independent. For any sequence of points $k_0 < k_1 < k_2$ along the vertical axis, $D_q^{-1}(k_j) - D_q^{-1}(k_{j-1})$ are statistically independent.

If
$$m_S = E\{S\} \quad \text{and} \quad \sigma_S^2 = \text{Var}\{S\} \tag{4.21}$$

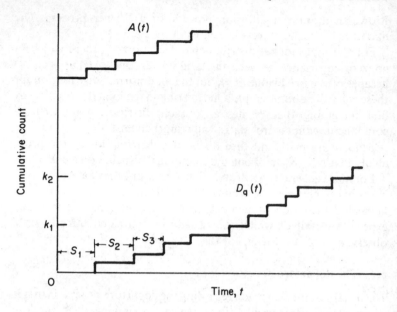

Figure 4.3   *Departure process with independent service times*

are finite then, according to the central limit theorem,

$$T_n = D_q^{-1}(n) = S_1 + S_2 + \ldots + S_n \qquad (4.22)$$

will have approximately a normal distribution for $n \gg 1$ with

$$E\{T_n\} = nm_S, \quad \text{Var}\{T_n\} = n\sigma_S^2, \quad \sigma_T = \sqrt{n}\,\sigma_S. \qquad (4.23)$$

More precisely,

$$\lim_{n \to \infty} P\left\{\frac{T_n - nm_S}{\sqrt{n}\,\sigma_S} < x\right\} = \frac{1}{(2\pi)^{1/2}} \int_{-\infty}^{x} \exp\left(-z^2/2\right) dz = \Phi(x).$$

$$(4.24)$$

For any sequence of integers $0 < k_1 < k_2 < \ldots$ with $k_j - k_{j-1} \gg 1$ $D_q^{-1}(k_j) - D_q^{-1}(k_{j-1})$ will be independent and approximately normal. Thus, the process T_n is a process of independent increments and approximately a diffusion process.

Unfortunately, most of the queue properties we wish to analyze involve the differences $A(t) - D_q(t)$, the queue length, or $D_q^{-1}(k) - A^{-1}(k)$, the waiting time (for FIFO). It is usually most convenient to describe the properties of A in terms of its vertical increments whereas it is easiest to describe D_q in terms of its horizontal increments. To subtract the two curves vertically or horizontally, however, it would be desirable to have the two processes described in a similar way. A homogeneous Poisson arrival process or a service with independent and exponentially distributed service times has the unique property that the A or D_q are (exactly) processes of independent increments both vertically and horizontally (which explains why these processes are so popular for mathematical analysis). The most general class of processes with this property is a homogeneous compound Poisson process with a geometric distribution of batch sizes, which is equivalent to a renewal process with times between events equal to zero with probability p but exponentially distributed with probability $1 - p$. This is approximately true, however, for more general processes; if $D_q^{-1}(k)$ or $A(t)$ is approximately a diffusion process so is $D_q(t)$ or $A^{-1}(k)$.

One. can construct a formal proof that $D_q(t)$ or $A^{-1}(k)$ is approximately a diffusion process from the fact that T_n is approximately normal and that $D_q(t)$ is a monotone nondecreasing function of t; the event $D_q(t) < n$ is, therefore, equivalent to the event $T_n > t$ (see Fig. 4.4). Intuitively, however, this is rather obvious from Fig. 4.4 which shows several possible realizations of $D_q(t)$ in the

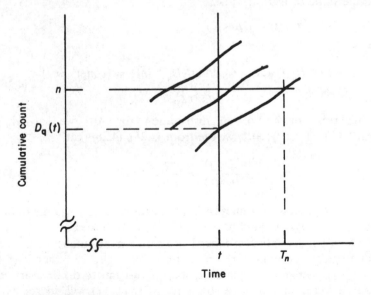

Figure 4.4 *Realizations of cumulative departure curves*

vicinity of some point in the (t, n) plane, with $n \gg 1$. The scale of Fig. 4.4 is so coarse, even locally, that one can disregard the discrete steps of $D_q(t)$; the scale of fluctuations is of order $n^{1/2}$.

The distribution of $D_q^{-1}(n)$ is the distribution of the time values where the various $D_q(t)$ curves cross the horizontal line of count n. The distribution is approximately normal according to (4.24). The distribution of $D_q(t)$ is the distribution of counts where the $D_q(t)$ curves cross the vertical line at time t. Most realizations $D_q(t)$, however, follow approximately a straight line of slope $1/m_S$. Thus, any sample point for $D_q(t)$ crossing a horizontal line maps approximately linearly into a sample point for the same curve crossing a vertical line. Since any linear function of a normally distributed random variable is also normally distributed, it follows that if $D_q^{-1}(n)$ or $A(t)$ is approximately normal so is $D_q(t)$ or $A^{-1}(n)$.

If in Fig. 4.4, for any specified value of n, we choose a t such that $t = E\{D_q^{-1}(n)\} = nm_S$, i.e., the vertical line at t passes through the mean of the normal distribution for $D_q^{-1}(n)$, then the mean of the distribution of $D_q(t)$ will be at n. Thus

$$E\{D_q(t)\} = n = t/m_S = t/E\{S\}. \tag{4.25}$$

The standard deviations of $D_q(t)$ and $D_q^{-1}(n)$ must have a ratio equal to the slope of $E\{D_q(t)\}$; i.e.,

$$\sigma_{D_q}(t) = (1/m_S)\sigma_{T_n},$$

thus

$$\text{Var}\{D_q(t)\} = (1/m_S)^2 \text{Var}\{D_q^{-1}(n)\} = (1/m_S)^2 (n\sigma_S^2)$$
$$= (t/m_S)(\sigma_S^2/m_S^2). \tag{4.26}$$

Whereas for a Poisson distribution for $A(t)$ we saw that $\text{Var}\{A(t)\}/E\{A(t)\} = 1$, we see from (4.25), (4.26) that

$$\frac{\text{Var}\{D_q(t)\}}{E\{D_q(t)\}} = \frac{\sigma_S^2}{m_S^2} = C^2(S). \tag{4.27}$$

For most processes of interest in applications $C^2(S)$ is comparable with 1 although it could have any values from 0 to ∞.

There is some statistical dependence between the number of service completions in non-overlapping time intervals, i.e., the $D_q(\tau_k) - D_q(\tau_{k-1})$ are not independent, because the time to the first service completion after time τ_k in the interval (τ_k, τ_{k+1}) will, in general, depend upon the time since the last completion in the previous interval (τ_{k-1}, τ_k). If the intervals (τ_{k-1}, τ_k) are large enough to include many services, however, the dependencies between the $D_q(\tau_k)$ $-D_q(\tau_{k-1})$ become insignificant.

4.8 Queue lengths and waiting times

Despite the fact that the arrival and service processes may have quite different detailed stochastic behavior, we have seen that, for many such processes, $A(\tau_k) - A(\tau_{k-1})$ and $A^{-1}(n_k) - A^{-1}(n_{k-1})$ are both approximately normal provided, in addition, that the server stays busy.

It is also true that if X and Y are two statistically independent normally distributed variables, then $X + Y$ or $X - Y$ (or $aX + bY$) are normally distributed. This is one of the key properties of normally distributed random variables and is, in part, one of the reasons for the existence of a central limit theorem. Clearly any special limit distribuion that might exist for arrivals in a time t must have the property that if it holds for times τ_1 and τ_2, it must hold also for $\tau_1 + \tau_2$. But the arrivals in time $\tau_1 + \tau_2$ is the sum of those in times τ_1 and τ_2. Thus the limit distribution must have the property that sums of random variables with this type distribution must also have this type distribution.

One immediate consequence of this is that if the number of arrivals in time t, $A(t) - A(0)$, and the number of departures in time t, $D_q(t) - D_q(0)$, are approximately normal and statistically independent, then so is their difference approximately normal. The difference, however, represents the change in queue length in time t, $Q(t) - Q(0)$, with

$$E\{Q(t) - Q(0)\} = E\{A(t) - A(0) - D_q(t) + D_q(0)\}$$
$$= E\{A(t) - A(0)\} - E\{D_q(t) - D_q(0)\},$$

$$\text{Var}\{Q(t) - Q(0)\} = \text{Var}\{A(t) - A(0) - D_q(t) + D_q(0)\}$$
$$= \text{Var}\{A(t) - A(0)\} + \text{Var}\{D_q(t) - D_q(0)\}.$$
$$(4.28)$$

One should notice that variances add while the expectations subtract. Also $D_q^{-1}(j) - A^{-1}(j)$ represents the waiting time of customer j if the queue discipline is FIFO. If the time to serve j customers $D_q^{-1}(j)$ $-D_q^{-1}(0)$ and the time for j customers to arrive $A^{-1}(j) - A^{-1}(0)$ are statistically independent, then the difference in waiting time between the jth and 0th customers will be approximately normal (for $j \gg 1$)

provided that the queue stays positive;

$$E\{w_j - w_0\} = E\{D_q^{-1}(j) - D_q^{-1}(0) - A^{-1}(j) + A^{-1}(0)\}$$
$$= E\{D_q^{-1}(j) - D_q^{-1}(0)\} - E\{A^{-1}(j) - A^{-1}(0)\}$$

and

$$\text{Var}\{w_j - w_0\} = \text{Var}\{D_q^{-1}(j) - D_q^{-1}(0)\}$$
$$+ \text{Var}\{A^1(j) - A^{-1}(0)\}. \qquad (4.29)$$

Suppose, for example, that at time 0 we have a queue $Q(0) \gg 1$ and that the average arrival and service rates are constants λ and μ respectively. We do not expect the number of services $D_q(t) - D_q(0)$ to depend upon the arrivals $A(t) - A(0)$ at least until such time when the queue might vanish and interrupt the server. During such time

$$E\{A(t) - A(0)\} = \lambda t, \quad E\{D_q(t) - D_q(0)\} = \mu t, \qquad (4.30)$$

and

$$E\{Q(t)\} = E\{Q(0)\} + (\lambda - \mu)t. \qquad (4.31)$$

The expected properties behave as in the fluid approximation; the average queue either increases or decreases accordingly as $\lambda > \mu$ or $\lambda < \mu$.

Suppose also that the arrival and departure processes are such that

$$\frac{\text{Var}\{A(t) - A(0)\}}{E\{A(t) - A(0)\}} = I_A, \quad \frac{\text{Var}\{D_q(t) - D_q(0)\}}{E\{D_q(t) - D_q(0)\}} = I_D \qquad (4.32)$$

are independent of time (at least for sufficiently large t that $E\{A(t) - A(0)\} \gg 1$). For example, if the arrivals are a Poisson process, then $I_A = 1$; and if the service times are independent, then $I_D = C^2(S)$ as in (4.27). For a given value of $Q(0)$

$$\text{Var}\{Q(t)\} = (I_A\lambda + I_D\mu)t, \qquad (4.33)$$

and

$$\sigma_Q = (I_A\lambda + I_D\mu)^{1/2}t^{1/2}$$

increases in proportion to $t^{1/2}$. If the initial queue is also considered to be a random variable but is independent of the subsequent arrivals and departures, then

$$\text{Var}\{Q(t)\} = \text{Var}\{Q(0)\} + (I_A\lambda + I_D\mu)t. \qquad (4.34)$$

To describe typical fluctuations in $Q(t)$ it is convenient to draw curves analogous to Figs 4.1 and 4.2 of $E\{Q(t)\} \pm \sigma_Q$ as illustrated in

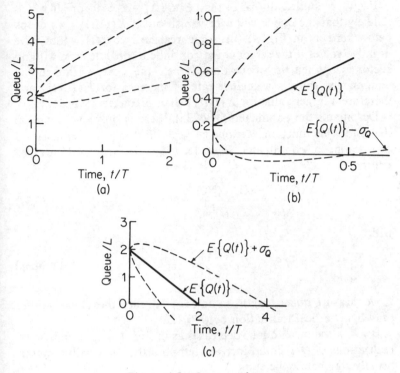

Figure 4.5 *Queue evolutions*

Fig. 4.5 by the broken lines. We expect that most realizations of $Q(t)$ will stay between these curves most of the time.

If $\lambda > \mu$ the curve

$$E\{Q(t)\} - \sigma_Q = Q(0) + (\lambda - \mu)t - (I_A\lambda + I_D\mu)^{1/2}t^{1/2} \quad (4.35)$$

will at first decrease with t since the σ_Q term will dominate the increasing mean for

$$(\lambda - \mu)t \ll (I_A\lambda + I_D\mu)^{1/2}t^{1/2}$$

i.e., for

$$t \ll \frac{(I_A\lambda + I_D\mu)}{(\lambda - \mu)^2} = T_0 \quad (4.36)$$

but for $t \gg T_0$ the opposite is true; the term $(\lambda - \mu)t$ becomes large compared with σ_Q; the queue is eventually almost certain to grow.

If $Q(0)$ is sufficiently large that $E\{Q(t)\} - \sigma_Q > 0$ for all t, it is unlikely that the queue will ever vanish but if $E\{Q(t)\} - \sigma_Q$ drops below zero as in Fig. 4.5(b), some realizations $Q(t)$ would have negative values if the server could continue to operate even with no customers. When the queue does vanish, the server is interrupted temporarily, causing a smaller rate of increase for $E\{D_q(t)\}$ and therefore a larger value for $E\{Q(t)\}$ than predicted by (4.31).

Despite all the parameters in (4.35), there is only one universal shape for this function. Various curves differ only by a translation or rescaling of coordinates. If we let $t^* = t/T_0$, then (4.35) can be written as

$$\frac{E\{Q(t)\} - \sigma_Q}{L_0} = \frac{Q(0)}{L_0} + t^{*1/2}(t^{*1/2} - 1) \qquad (4.37)$$

with

$$L_0 = \frac{I_A \lambda + I_D \mu}{|\lambda - \mu|}. \qquad (4.38)$$

If we measure queue lengths in units of L_0, then these curves differ only by a vertical translation by $Q(0)/L_0$.

If $\lambda < \mu$ the mean queue decreases as in Fig. 4.5(c) but individual realizations of $Q(t)$ could increase temporarily. They will, however, usually stay below the curve

$$E\{Q(t)\} + \sigma_Q = Q(0) - (\mu - \lambda)t + (I_A \lambda + I_D \mu)^{1/2} t^{1/2}$$
$$= Q(0) - L_0 t^{*1/2}(t^{*1/2} - 1)$$

until the queue vanishes.

A similar description can be made of the evolution of waiting times w_j. If customer 0 must wait a time w_0 before entering the service, the analogues of (4.30), (4.31) would be

$$E\{A^{-1}(j) - A^{-1}(0)\} = j/\lambda, \quad E\{D_q^{-1}(j) - D_q^{-1}(0)\} = j/\mu$$

and

$$E\{w_j\} = E\{w_0\} + j(1/\mu - 1/\lambda). \qquad (4.39)$$

The analogues of assumptions (4.32) would be that

$$\frac{\text{Var}\{A^{-1}(j) - A^{-1}(0)\}}{E\{A^{-1}(j) - A^{-1}(0)\}} = \frac{I_A}{\lambda}, \quad \frac{\text{Var}\{D_q^{-1}(j) - D_q^{-1}(0)\}}{E\{D_q^{-1}(j) - D_q^{-1}(0)\}} = \frac{I_D}{\mu}$$
$$(4.40)$$

and so

$$\text{Var}\{w_j\} = \left(\frac{I_A}{\lambda^2} + \frac{I_D}{\mu^2}\right)j \qquad (4.41)$$

$$\sigma_w = \left(\frac{I_A}{\lambda^2} + \frac{I_D}{\mu^2}\right)^{1/2} j^{1/2}. \qquad (4.42)$$

Typical fluctuations of the w_j will lie between curves

$$E\{w_j\} \pm \sigma_w = w_0 + (1/\lambda - 1/\mu)j \pm \left(\frac{I_A}{\lambda^2} + \frac{I_D}{\mu^2}\right)^{1/2} j^{1/2} \quad (4.43)$$

which can be rescaled to have the forms

$$\frac{E\{w_j\} \pm \sigma_w}{W_0} = \frac{w_0}{W_0} \pm j^{*1/2}(j^{*1/2} \pm 1). \qquad (4.44)$$

with $j^* = j/N_0$ and

$$N_0 = \frac{I_D + I_A\mu^2/\lambda^2}{(1 - \mu/\lambda)^2}, \quad W_0 = \frac{I_D + I_A\mu^2/\lambda^2}{\mu(1 - \mu/\lambda)}. \qquad (4.45)$$

The above formulas are obviously useful to the description of the queue behavior if $\lambda > \mu$ and $Q(0)$ or w_0 are so large that the queue is not likely to vanish. Under such conditions the queue length and waiting times should be approximately normally distributed with the appropriate means and variances. Even if, for $\lambda > \mu$, the queue is likely to vanish some time (particularly if $Q(0) = 0$), by time $t \simeq T_0$ the queue is likely to be large enough (comparable with L_0) that it will not vanish again for $t \gtrsim T_0$. If we consider the queue distribution at $t = T_0$ as a new initial state, then for $t \gg T_0$ the mean and variance of $Q(t) - Q(T_0)$ will be large compared with $E\{Q(T_0)\}$ and $\text{Var}\{Q(T_0)\}$. We still conclude that $Q(T_0)$ will be approximately normally distributed with slight corrections to the mean and variance due to the modified initial conditions. One can also make obvious extensions of these formulas if $\lambda(t)$ or even $\mu(t)$ are not constant. What happens for $\lambda < \mu$, after the queue vanishes, will be discussed later.

Several people have given rigorous mathematical proofs that, for queueing models of specific type, the queue or waiting time distributions will have the limiting normal form when suitably rescaled, if $\lambda > \mu$. Any exact formulas for these distributions, however, are so complicated as to be virtually useless even for the simplest types of arrival or service processes. Fortunately, there are no obvious applications of queueing theory for which one needs to evaluate queue lengths or waits to an accuracy of better than 10% or so. Properties of the arrival or service processes are not usually known well enough to justify calculations of greater precision.

4.9   Work conserving systems

To obtain some crude description of how a system behaves one might choose actually to draw some possible realizations $A^{(j)}(t)$ as discussed in Section 4.1. If the server is a single- or multiple-channel server with independent service times, one could generate a sequence of random numbers with a distribution $F_S(t)$, construct one or more possible departure curves for each $A^{(j)}(t)$, and measure whatever one wishes from these realizations.

For a work conserving server, however, it may be easier to construct realizations of the cumulative arrivals and departures of work $A^W(t)$, $D^W(t)$. Distances on these curves do not describe quite the same information as on the curves for customer counts, but it generally provides just as useful a description of the system behavior. The work curves have the advantage that most of the stochastic fluctuations appear on the $A^W(t)$ curves. Whenever there is a queue of work, the $D^W(t)$ curve for a single-channel server or each component of a multiple-channel server behaves as in Fig. 3.10. The height and width of successive steps may be different, but the curve never deviates by more than one (random) service time from a line of slope 1. If one is interested in the individual steps because the relevant queues involve only a few customers (possibly only 1 or 2), one can draw them; but if the typical queue involves several or many customers, one might approximate the $D^W(t)$ curve by a straight line of slope 1 for a single-channel server or of slope m for an m-channel server (when there is a queue).

For a single-channel server a line of slope 1 diagonally through the corners of the squares between D_q^W and D_s^W in Fig. 3.10 has a special significance itself. At any time t the vertical distance from this diagonal up to D_q^W represents the residual service time of the customer already in service. The vertical distance from this line to $A^W(t)$ thus represents the work in queue plus the residual service of the customer in the server. In the queueing literature, this is often called the 'virtual work'. If the queue discipline is FIFO, it represents the time a hypothetical customer who arrives at time t would wait to enter the service.

The work to arrive during some period of time, say from 0 to t, can be represented as the sum of the service times of all customers to arrive during this time, i.e.,

$$A^W(t) - A^W(0) = \sum_{j=A(0)+1}^{A(t)} S_j. \qquad (4.46)$$

If the number of customers is large compared with 1, $A^W(t) - A^W(0)$ should be approximately normal with

$$E\{A^W(t) - A^W(0)\} = E\{S\} E\{A(t) - A(0)\} \qquad (4.47)$$

and

$$\text{Var}\{A^W(t) - A^W(0)\} = E\{A(t) - A(0)\} \text{Var}\{S\}$$
$$+ \text{Var}\{A(t) - A(0)\} E^2\{S\}. \qquad (4.48)$$

If, in addition, the variance to mean ratio for the arrivals is (approximately) independent of time as in (4.32), then

$$\text{Var}\{A^W(t) - A^W(0)\} = E\{A(t) - A(0)\}[\text{Var}\{S\} + I_A E^2\{S\}]$$
$$= E^2\{S\} E\{A(t) - A(0)\}[I_D + I_A].$$

Thus

$$\frac{\text{Var}\{A^W(t) - A^W(0)\}}{E\{A^W(t) - A^W(0)\}} = E\{S\}[I_D + I_A] \qquad (4.49)$$

is (nearly) independent of t.

If during a time t the queue does not vanish, $D^W(t)$ will be linear and have no variance. Thus $\text{Var}\{Q^W(t) - Q^W(0)\}$ is equal to $\text{Var}\{A^W(t) - A^W(0)\}$.

If (4.49) is true and queues are expected to be fairly large (say 5 or more) most of the time, one can obtain a coarse graphical representation of the queue behavior by first drawing a graph of $E\{A^W(t)\}$, assumed to be known. Next, generate some hypothetical (approximate) realizations of $A^W(t)$ by adding to $E\{A^W(t)\}$ realizations of a suitably scaled Brownian motion of zero mean. Finally, for each realization of $A^W(t)$, one constructs the $D_q{}^W(t)$ by drawing a curve having slope 1 everywhere that $D^W(t) < A^W(t)$, and measures from the graphs whatever is of interest.

To generate the appropriate Brownian motion one can start with a set of random samples X_i from any convenient distribution with known variance $\sigma_x < \infty$ and $E\{X\} = 0$; for example, $X_i = +1$ or -1 with probability 1/2. For each of several independent sequences of the X_i draw graphs of

$$Z(n) = \sum_{i=1}^{n} X_i$$

as a function of n.

These realizations of $Z(n)$ satisfy

$$E\{Z(n)\} = 0, \quad \text{Var}\{Z(n)\} = n\sigma_x^2, \quad \sigma_{Z(n)} = \sqrt{n}\,\sigma_x.$$

By rescaling this with any constant $\alpha > 0$, one also generates realizations of $\alpha Z(n)$ having a variance $n\alpha^2 \sigma_x^2$, in particular

$$Z^*(n) = \frac{E\{S\}(I_D + I_A)^{1/2}}{\sigma_x} Z(n) \qquad (4.50)$$

has a variance

$$\text{Var}\{Z^*(n)\} = E^2\{S\}(I_D + I_A)n.$$

To construct an approximate realization of $A^W(t) - E\{A^W(t)\}$, at a suitable set of discrete times t_j evaluate the dimensionless number

$$E\{A^W(t_j) - A^W(0)\}/E\{S\} \qquad (4.51)$$

from the graph of $E\{A^W(t_j)\}$ (this number should be moderately large compared with 1). With this as the value of n, let

$$A^W(t_j) - A^W(0) = Z^*(n).$$

This construction can be done quite rapidly with one or two proportional dividers as illustrated in Fig. 4.6. First, draw the curve $E\{A^W(t)\}$ and several realizations $Z(n)$ (two are shown in the figure). To evaluate (4.51), measure the distance $E\{A^W(t) - A^W(0)\}$ on Fig. 4.6 with proportional dividers set with a ratio $1/E\{S\}$ and transfer the scaled distance as n. Equivalently, one can draw a line A–B of slope $1/E\{S\}$ through $E\{A(0)\}$ and for each t_j draw a horizontal line B–C at height $E\{A^W(t_j)\}$. Project the point B down to the $Z(n)$ graph as n. With (another) proportional dividers set with a ratio $(I_D + I_A)^{1/2}E\{S\}/\sigma_x$, measure the distance $b' = Z(n)$ and transfer the scaling of this distance as $b = A^W(t_j) - E\{A^W(t_j)\}$. Although this must be done for many values of t_j to generate an approximate realization $A(t)$, for each t_j one can evaluate the $A(t_j)$ for several $Z(n)$ realizations together.

Problems

4.1 If in problem 1.1 the Y_i and Y_i' are independent random variables with $p = P\{Y_i = 1\} = 0.4$ and $p' = P\{Y_i' = 1\} = 0.5$, what is the probability that the time interval between two arrivals has the value $k = 1, 2, \ldots$ if the arrivals are generated by (a) the Y_i, (b) the Y_i'?

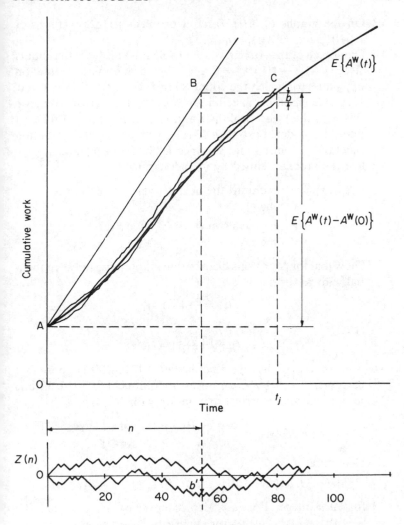

Figure 4.6   *Construction of queue realizations*

4.2 From each of the two $A(t)$ curves constructed in problem 1.1, evaluate the number of times, n_k, that the time interval between arrivals has the value $k = 1, 2, \ldots$. Draw a histogram, i.e., a graph of n_k versus k. From the probabilities in problem 4.1 and the actual number of arrivals, $\Sigma_k n_k$, determine $E\{n_k\}$ and $\text{Var}\{n_k\}$. Compare the observed histogram with graphs of $E\{n_k\} \pm \sigma_{n_k}$.

4.3 On the graphs of $A(t)$, $D_s(t)$ in problem 1.1 draw curves of $E\{A(t)\}$ and $E\{A(t)\} \pm \sigma_{A(t)}$.

4.4 From each of the $A(t)$ curves in problem 1.1 evaluate the counts $A(15(j+1)) - A(15j), j = 0, 1, \ldots 9$ and determine the fraction of j values for which the count is less than or equal to $n, n = 0, 1, \ldots$ (an empirical distribution function for $A(15)$). Compare this with the theoretical distribution function for $A(15)$ based upon the model of problem 4.1, and with a normal distribution function with mean and variance $E\{A(15)\}$, $\mathrm{Var}\{A(15)\}$.

4.5 For the process defined by (4.5) determine

$$P_0(t, \tau) = P\{\text{time until the next arrival} > \tau | \text{an arrival at time } t\}$$
$$= P\{\text{no arrivals in } (t, t+\tau) | \text{an arrival at time } t\}.$$

Show that for $N \gg 1$ this distribution is approximately exponential with parameter

$$\lambda(t) = \frac{\mathrm{d}}{\mathrm{d}t} E\{A(t)\}$$

i.e.,

$$P_0(t, \tau) \simeq \exp(-\lambda(t)\tau).$$

4.6 For a process of the type described in (4.13), (4.14), suppose that customer j, $-\infty < j < \infty$, has an appointment to arrive at time $j\delta$ but he actually arrives at time T_j^* with

$$F_j(t) = \begin{cases} 1 - \exp(-\lambda(t-j\delta)) & \text{for } t \geq j\delta \\ 0 & \text{for } t < j\delta. \end{cases}$$

Determine

$$E\{A(t) - A(0)\} \quad \text{and} \quad \mathrm{Var}\{A(t) - A(0)\}$$

for t/δ = integer. For $\lambda\delta \ll 1$ consider the behavior of $\mathrm{Var}\{A(t) - A(0)\}/E\{A(t) - A(0)\}$ for small and large values of t/δ.

4.7 If, for the processes discussed in Section 4.8, $Q(0)$ is sufficiently large and $\lambda < \mu$, show that the time until the queue vanishes for the first time is approximately normally distributed.

4.8 Three airplanes are scheduled to arrive at an airport at the same time $t = 0$. The actual times of arrival, however, are T_1^*, T_2^*, and T_3^*, independent and identically distributed random variables with a distribution function

$$F_T(t) = P\{T_j^* < t\} = \begin{cases} 1 - e^{-t/\tau} & t > 0 \\ 0 & t < 0 \end{cases}$$

If there is a minimum time β between each landing, what is the probability that (a) there is never a queue, (b) there is ever a queue of two?

4.9 The time interval H_k between consecutive arrivals $k - 1$ and k are independent and identically distributed (i.i.d.) with a distribution

$$P\{H_k > h\} = 1 - F_H(h) = \begin{cases} 1 & \text{for} \quad h < 0 \\ (1-p)e^{-\lambda h} & \text{for} \quad h \geq 0 \end{cases}$$
$$0 < p < 1.$$

Show that this arrival process is equivalent to a Poisson process of i.i.d. batches B_j arriving at a rate λ. Evaluate the distribution of B_j, $p_j = P\{B = j\}$. If $A(t)$ is the number of arrivals in time t determine

$$I_A = \frac{\text{Var}\{A(t)\}}{E\{A(t)\}} = \frac{\text{Var}\{H\}}{E^2\{H\}} = \frac{E\{B^2\}}{E\{B\}}.$$

as a function of p and λ.

4.10 Cars queue at a toll plaza having n toll booths. At the kth toll booth cars can pass with a time headway of S_{jk} seconds between the jth and $(j+1)$th car. The S_{jk} are statistically independent random variables with

$$E\{S_{jk}\} = m_S, \quad \text{Var}\{S_{jk}\} = \sigma_S^2, \quad j = 1, 2, \ldots$$
$$k = 1, 2, \ldots, n.$$

For a time t large compared with m_s, evaluate

$$\text{Var}\{D_q(t)\}/E\{D_q(t)\}$$

in which $D_q(t)$ is the cumulative total number of cars to leave the toll plaza from all booths.

4.11 Suppose we have a single-channel queueing system, for which

$$\frac{\mathrm{d}}{\mathrm{d}t} E\{Q(t)|Q(t) > 0\} = \lambda - \mu > 0,$$

$$\frac{\mathrm{d}}{\mathrm{d}t} \text{Var}\{Q(t)|Q(t) > 0\} = I_A + I_D.$$

i.e., the average queue grows at a rate $\lambda - \mu$ and the variance of the queue grows at a rate $I_A + I_D$ whenever the queue is positive. The system starts at time $t = 0$ with $Q(0) = 0$. Describe why, for sufficiently large t,

$$E\{Q(t)\} - (\lambda - \mu)t$$

should approach some constant $C > 0$. Estimate the order of magnitude of C and the order of magnitude of the time t required for this quantity to reach approximately the value C.

4.12 Taxis arrive at a loading platform according to a homogeneous Poisson process with the rate λ_T. Passengers arrive according to a homogeneous Poisson process with rate λ_p. Passengers instantaneously board a taxi if one is available (one person per taxi) and loaded taxis depart immediately. At any ime there may be either a 'taxi queue' or a 'passaeger queue' but never both simultaneously. At time zero there is neither a taxi nor a passenger queue.

Describe the probability distribution of the taxi or passenger queue at time t, if $\lambda_T t$ and $\lambda_p t$ are both large compared with 1 (about 100, for example). The description should include cases $\lambda_T > \lambda_p$, $\lambda_T = \lambda_p$, or $\lambda_T < \lambda_p$.

CHAPTER 5

Equilibrium distributions

5.1 Stationary processes

In Chapter 4 we saw approximately how queues evolve if they do not vanish. For $\lambda > \mu$ the mean queue simply follows what the fluid approximation predicts, but the uncertainty in the queue length as measured by its standard deviation typically grows as the square root of the cumulative arrivals and departures. If, however, we start with an initial queue $Q(0) > 0$ and λ is less than μ, the mean queue decreases as illustrated in Fig. 4.5(c). The queue is certain to vanish within a finite time, usually within a time comparable with $Q(0)/(\mu - \lambda)$.

After the queue vanishes for the first time, it may stay zero for a while, but sooner or later a customer will arrive while another is being served causing a queue of length one. It is also possible during any given period of time that the actual number of arrivals exceeds the actual number of departures by considerably more than one, even though $\mu > \lambda$. Once a queue forms, it will on the average decrease, but for any particular realization, fluctuations may cause it to increase temporarily. For a typical realization of $Q(t)$, the queue will vanish, maybe stay zero for a while, reform, vanish again, etc.

The length a queue will reach during one of these fluctuations depends upon the relative size of the variance in arrivals and departures which cause the temporary queue, and the 'excess capacity' $\mu - \lambda$ which measures the expected rate at which the queue decreases once it has formed. If $\mu - \lambda$ is small, the 'restoring force' that tries to destroy the queue is weak and fluctuations may cause very sizeable queues to form.

If we interpret $Q(t)$ as a random function, then for any times $\tau_1, \tau_2 \ldots$, the queue lengths $Q(\tau_1)$, $Q(\tau_2)$, $\ldots$ represent a set of random variables with some joint probability distribution. For any specified t, $Q(t)$ is random variable with a (marginal) distribution function

$$F_Q(x; t) = P\{Q(t) \le x\}.$$

As always, the physical interpretation of this is that one, somehow, has a means of repeating the observation of the queue evolution under 'identical conditions'. If the experiment were repeated with the same initial conditions $Q(0)$ (and anything else that may be relevant), then $F_Q(x; t)$ represents the fraction of repetitions for which the queue at time t has a value less than or equal to x. Although one may never, in fact, do this, this is the 'thought experiment' that one should associated with these distributions.

Most of the literature on queueing theory deals with the stochastic properties of queue lengths, waiting times, etc., when the arrivals form a stationary process and the properties of the service are time independent.

A stochastic arrival process is defined (mathematically) to be stationary if the numbers of arrivals $A(\tau_k) - A(\tau_{k-1})$, $k = 1, 2, \ldots n$ during any time intervals (τ_0, τ_1), (τ_1, τ_2), $\ldots$, (τ_{n-1}, τ_n) have the same joint probability distribution as the numbers of arrivals $A(\tau_k + \tau) - A(\tau_{k-1} + \tau)$ during the time intervals $(\tau_0 + \tau, \tau_1 + \tau)$, $(\tau_1 + \tau, \tau_2 + \tau)$ $\ldots$ for all values of τ and any choice of the τ_k and n. Thus, all stochastic properties of the arrivals are invariant under a translation by any time τ; the process looks the same no matter when one starts to observe it. For a stationary arrival process the cumulative function $A(t)$ is described as a random function of stationary increments. A homogeneous Poisson process, for example, is a stationary process.

For a time independent server, the service times have probability distributions which may depend upon the type of customer, past services performed or other 'state variables' which may vary with time, but the time to perform a given sequence of tasks does not depend upon when it is done. If the tasks to be performed are invariant under translations in time and the server is kept busy, presumably the departure process $D_q(t)$ is also a process of stationary increments. For a work conserving system, the arrival process of work would be considered stationary if the $A^w(t)$ is also a random function with stationary increments.

For any stationary arrival process, the distribution of $A(t + \tau) - A(\tau)$ must, in particular, be independent of τ, consequently $E\{A(t + \tau) - A(\tau)\}$ is also independent of τ. Since the expected number of arrivals in $(t, t + \tau)$ is the sum of the expected arrivals in any partition of this interval, it can be written as

$$E\{A(t+\tau)-A(\tau)\} = \int_{\tau}^{t+\tau} \lambda(t')dt' \quad \text{with} \quad \lambda(t) = \frac{dE\{A(t)\}}{dt}.$$

That this is independent of τ implies that $\lambda(t) = \lambda$ is independent of t and

$$E\{A(t+\tau)-A(\tau)\} = \lambda t.$$

Similarly, if a time-independent server is kept busy

$$E\{D_q(t+\tau)-D_q(\tau)\} = \mu t,$$

and for a work conserving system with stationary arrivals

$$E\{A^w(t+\tau)-A^w(\tau)\} = E\{S\}\lambda t$$

independent of τ.

For stationary arrivals $\text{Var}\{A(t+\tau)-A(\tau)\}$ is independent of τ as is

$$\frac{\text{Var}\{A(t+\tau)-A(\tau)\}}{E\{A(t+\tau)-A(\tau)\}} = I_A(t).$$

If there is zero covariance between the number of arrivals in non-overlapping time intervals, then the variance of the arrivals in $(t, t+\tau)$ would be equal to the sum of the variances of the arrivals in any partition of $(t, t+\tau)$ and we would conclude (as with the expectations) that $\text{Var}\{A(t+\tau)-A(\tau)\}$ is proportional to t and $I_A(t)$ is independent of both t and τ. Such would be the case if $A(t)$ is a process of independent increments.

A stationary process is, of course, a mathematical abstraction. No arrival process encountered in queueing applications is really stationary; most such processes have rush hours, weekly variations, seasonal variations, etc., as discussed in Section 2.1. To use a stochastic model of any kind one must imagine that an arrival process which may continue for years is first partitioned into arrivals during 24-hour (or weekly) periods and then further stratified into what are considered as similar days (weeks). What happens on similar days are then interpreted as the repetitions 'under identical conditions'.

Within these 24-hour periods there may be certain time intervals (possibly of 30 minutes or so duration) during which $\lambda(t)$, as determined from the average of the $A^{(j)}(t)$ over similar days, is 'nearly' constant. It is not clear yet, of course, in what sense one can approximate a slowly varying $\lambda(t)$ by a constant which is actually treated as a parameter whose value changes (slowly) with time. In applications, this problem is further complicated by the fact that there may also be trends in what were interpreted as equivalent days, i.e., the process may not be quite stationary over the days either. We are interested here in the behavior of queues, delays, etc., for arrival processes that are nearly stationary, and for the hypothetical situation in which they are exactly stationary.

Suppose that at time 0 we start with some arbitrary initial queue $Q(0)$, and for $t > 0$ the arrivals are stationary, the service is time independent and $\lambda < \mu$. One would expect, for most reasonable processes, that the probability distribution of $Q(t)$ would eventually approach a limit distribution which does not depend upon $Q(0)$, i.e.,

$$F_Q(x; t | Q(0)) = P\{Q(t) \le x | Q(0)\} \to F_Q(x) \quad \text{for} \quad t \to \infty \quad (5.1)$$

with $F_Q(x)$ independent of t and $Q(0)$. Even if $Q(0)$ is initially very large, the condition $\lambda < \mu$ should force the queue to vanish within a finite time and eventually to follow a behavior determined by fluctuations in the stationary arrival and service processes.

This behavior is not guaranteed by the postulates of a stationary process for arrivals and service and, in fact, is true only for a rather special class of processes, primarily those in which the fluctuations during time intervals sufficiently separated from each other are nearly statistically independent. In the probability literature such processes are described as 'ergodic' or as satisfying a 'mixing condition'. We will not worry here about the precise mathematical definitions since it would be impossible to verify, in any finite time, whether or not a real process satisfied any precise conditions.

A classic example of an arrival process that is not ergodic is a process of the type discussed in Section 4.5 for which each realization behaves nicely (for example, like a homogeneous Poisson process), but for which the λ is itself interpreted as a random variable with some given expectation. Generally, if one has any family of stationary processes, a process generated by a random selection from this family is itself a stationary process. Although for given λ, the queue distribution may have a limit for $t \to \infty$ as in (5.1), this distribution will depend upon λ. The (unconditional) queue distribution at time t may, in turn, depend upon $Q(0)$ in that the value of λ or its distribution may depend upon $Q(0)$.

This example of a non-ergodic process is important in applications because many arrival processes do behave as if the λ observed on different days are different, and appear as if they were selected at random from some distribution. If such is the case, one should treat separately those processes with the same λ and hope that they behave as if they were ergodic. Any derived properties are then eventually averaged over the distribution of λ.

If, for sufficiently large t, the stochastic properties of $Q(t)$ do not depend upon $Q(0)$, then for any times $\tau_1 < \tau_2 < \ldots \tau_n$ with $\tau_k - \tau_{k-1}$ sufficiently large, the $Q(\tau_k)$ will be (nearly) statistically independent. One need only consider each τ_k as a new time origin and note that the conditional distributions of $Q(\tau_{k+1}), Q(\tau_{k+2}), \ldots$ given $Q(\tau_k)$ are, by hypothesis, independent of $Q(\tau_k)$.

This has important practical consequences. The experimental interpretation of $F_Q(x; t)$ is based upon 'repetition of the experiment under identical conditions' i.e., observing the fraction of realizations of $Q(t)$ for which $Q(t) < x$ at corresponding values of t. If, however, $Q(t)$ behaves as postulated, $F_Q(x; t)$ will have the equilibrium queue distribution $F_Q(x)$ for all sufficiently large t. To measure $F_Q(x)$, it suffices to observe a *single* realization $Q(t)$ at times $\tau_1, \tau_2, \ldots$ with sufficiently large $\tau_k - \tau_{k-1}$ and interpret the $Q(\tau_k)$ as independent samples from the distribution $F_Q(x)$. Thus

$$F_Q(x) \simeq n^{-1} \text{ (number of } k \text{ with } Q(\tau_k) < x)$$
$$= \text{fraction of } \tau_k \text{ for which } Q(\tau_k) < x. \qquad (5.2)$$

If someone else observed the same $Q(t)$ but at other times $\tau_1', \tau_2', \ldots, \tau_n'$, presumably he would obtain nearly the same estimate of $F_Q(x)$. If we combine the results of the two experiments, we would also conclude that

$$F_Q(x) \simeq (2n)^{-1} \text{ (number of } \tau_k \text{ or } \tau_l' \text{ for which } Q(t) < x)$$

even though the τ_k may not be well separated from the τ_l'.

By extending this argument to any number of collections of time points τ_k'', etc., it follows that the validity of (5.2) does not depend upon the τ_j being well separated as long as they are more or less evenly distributed over a long enough total time. Furthermore, one could take a continuum of observations and conclude that

$$F_Q(x) \simeq \text{fraction of time that } Q(\tau) < x. \qquad (5.3)$$

This can be evaluated easily from a graph of $Q(t)$. Draw a horizontal line at height x and measure the total length of time that the graph $Q(t)$ is below the horizontal line during some large but finite time period.

Equation (5.3) is interpreted to mean that the probability of finding $Q(t) < x$ at any time t if the queue distribution is stationary is equal to the long time average fraction of time that $Q(t) < x$. It follows also from (5.3) that, for any function $g(\cdot)$ the expectation of $g(Q)$ with Q distributed as $F_Q(x)$ is equal to the long time average of $g(Q(t))$, i.e.,

$$E\{g(Q)\} = \lim_{T \to \infty} \frac{1}{T} \int_0^T g(Q(\tau))\,d\tau, \tag{5.4}$$

Generally, properties of this type in which expectations over some probability distribution are related to time averages of a single time series are called ergodic properties.[†]

If $Q(t)$ is integer valued, (5.3) also implies that the probability of having $Q(t) = j$ at a single time t, when $Q(t)$ is in its equilibrium distribution, is equal to the fraction of time that any realization $Q(t)$ spends in the state j.

This notion of ergodicity can also be extended to joint distributions of queue lengths at two or more times. For example, $P\{Q(\tau) < x$ and $Q(\tau + t) < y\}$ represents the joint probability distribution of queue lengths at times τ and $\tau + t$, which in equilibrium should be independent of τ. If one observes queues at times τ_k and $\tau_k + t$ for fixed t but many τ_k, this joint probability should (for an ergodic system) be equal to the fraction of the τ_k for which $Q(\tau_k) < x$ and $Q(\tau_k + t) < y$. If one observes $Q(\tau)$ and $Q(\tau + t)$ at all times τ, the joint probability will be equal to the long time fraction of τ values for which $Q(\tau) < x$ and $Q(\tau + t) < y$. Also for any function $g(\cdot)$,

$$E\{g(Q(\tau), Q(\tau + t))\} = \lim_{T \to \infty} \frac{1}{T} \int_0^T g(Q(\tau'), Q(\tau' + t))\,d\tau'$$

independent of τ. For example,

$$\text{Cov}\{Q(\tau), Q(\tau + t)\} = E\{Q(\tau)Q(\tau + t)\} - E^2\{Q(\tau')\}$$

$$= \lim_{T \to \infty} \frac{1}{T} \int_0^T [Q(\tau')Q(\tau' + t) - E^2 Q(\tau')\}]\,d\tau'. \tag{5.4a}$$

---

† The term ergodic was coined from Greek words meaning work (energy) and path. It originally related to a hypothesis in statistical mechanics that the time average behavior of a molecular system over some path in a suitable state space of very large dimension was equal to the average over a surface of constant energy in this state space. The term has since been adopted to refer to any relation of the type (5.4) even though it may have no connection with energy or paths.

If in (5.2) the τ_k are sufficiently separated that the $Q(\tau_k)$ are statistically independent, then the right hand side of (5.2) is a random variable representing the fraction of successes in n independent trials when the probability of success is $F_Q(x)$. The number of successes has a binomial distribution with a standard deviation proportional to $n^{1/2}$. The fraction of successes has a standard deviation proportional to $n^{-1/2}$. This all relates, of course, to the classic problem in statistics of estimating the probability of success in Bernoulli trials.

If the $Q(\tau_k)$ are not statistically independent, (5.2) still holds but one cannot immediately infer the accuracy of the estimate. If, for example, one doubles the number of τ_k by taking observations at times τ'_l, but with $\tau_k = \tau'_k$, one has obtained no new information and has not improved the estimate at all. But if the τ_k and τ'_l are separated enough that the $Q(\tau_k)$ and $Q(t'_l)$ are all statistically independent, then the variance of the estimator is reduced by a factor of $1/2$. In the limit of observations at infinitely many times τ as in (5.3) or (5.4) it does not follow that for any finite time period the right hand side of (5.3) or (5.4) has zero variance as would be the case if the $Q(\tau)$ were statistically independent for all τ. The variance of these estimators can be evaluated in terms of the covariance (5.4a). The variance should certainly be less than for (5.2) but typically not by a large factor.

5.2 Dimensional estimates

It is possible to evaluate equilibrium queue distributions for certain special types of queueing models but the derivations of these distributions are rather formal and usually devoid of 'physical' intuition. Some are also quite complex. To obtain at least order of magnitude estimates of typical queue behavior, however, requires only elementary reasoning.

In Section 4.8 we saw that if the variance to mean ratio for arrivals and departures were (nearly) independent of time as in (4.32), a typical queue evolution would usually stay within the range

$$E\{Q(t)\} \pm \sigma_Q = Q(0) - (\mu - \lambda)t \pm (I_A \lambda + I_D \mu)^{1/2} t^{1/2}$$
$$= Q(0) - L_0 t^{*1/2}(t^{*1/2} \mp 1)$$

at least until the queue vanishes, with $t^* = t/T_0$ and L_0, T_0 defined in (4.36), (4.38). For the upper sign, this has a maximum of $Q(0) + L_0/4$ at $t^* = 1/4$.

Even if the initial queue is zero there is a significant probability that a queue will form immediately (or within a time comparable with $1/\mu$) and subsequently grow to a magnitude of order L_0 and last for a time of order T_0. If the queue vanishes at a later time (say within a time comparable with T_0), we can apply the argument again with this as a new starting time.

If a typical queue length is so large ($L_0 \gg 1$) that its integer nature becomes irrelevant, then there must be some natural scale with which it is to be measured. Although we have not identified the number L_0 with any specific quantitative property of the queue length (such as $E\{Q\}$), it clearly represents a natural scale for measuring any property of Q.

For $\lambda < \mu$ we can also write (4.38) as

$$L_0 = \frac{(I_D + I_A \rho)}{1 - \rho}, \quad \rho = \lambda/\mu. \tag{5.5}$$

Since for most processes encountered in applications I_A and I_D are comparable with 1, a simple rule of thumb is that queue lengths will typically be comparable with $1/(1 - \rho)$. If ρ is not close to 1, say $\rho < 1/2$, typical queue lengths will be only 1 or 2; the scale L_0 will be comparable with the scale of integers. Indeed, for a single-channel server we already know that the server must be idle a fraction $1 - \rho$ of the time. Thus, if $Q_s(t)$ represents the number of customers in the system (server plus queue), the ergodic property implies that $P\{Q_s(t) = 0\} = 1 - \rho$. On the other hand, if ρ is close to 1, L_0 will be large compared with 1; for $\rho \to 1, L_0 \to \infty$. For $\rho = 0.9, L_0 \simeq 10$; for $\rho = 0.99, L_0 \simeq 100$.

The time T_0 is the natural unit of time for observing any significant changes in $Q(t)$. It is a scale for measuring the length of time it takes for any initial queue of length comparable with L_0 to reach 0 and to reform with a value nearly independent of its initial value. Thus in the preceding section, the meaning of the τ_k being 'sufficiently separated' that the $Q(\tau_k)$ are (nearly) statistically independent is that the $\tau_{k+1} - \tau_k$ are comparable with T_0. If one wishes to estimate $F_Q(x)$ by observing $Q(t)$ over some extended period of time, then the standard deviation of the estimators (5.2) or (5.3) should be comparable with the $1/\sqrt{n}$ in (5.2). To obtain a fractional error of $1/\sqrt{n}$ in the estimate of $F_Q(x)$, one must observe $Q(t)$ for a time of order nT_0 (for a 10% accuracy $1/\sqrt{n} = 0.1$, one observes $Q(t)$ for a time comparable with $100T_0$, for a 1% accuracy one must observe $Q(t)$ for about $10^4 T_0$).

We can also write (4.36) in the form

$$T_0 = \frac{(I_D + I_A \rho)}{(1 - \rho)^2}(1/\mu) \qquad (5.6)$$

in which $1/\mu$ is the mean time between services when the server is busy; the mean service time for a single-channel server. A simple rule of thumb for most typical systems is that T_0 is comparable with $(1 - \rho)^{-2}(1/\mu)$.

If ρ is not close to 1 ($\rho < 1/2$), then T_0 is comparable with $1/\mu$; the system forgets its past within a time comparable with one service time. As ρ approaches 1, however, T_0 increases very rapidly; for $\rho = 0.9$, T_0 is about 100 $(1/\mu)$ and for $\rho = 0.99$ it is about 10^4 $(1/\mu)$. If one wishes to estimate $F_Q(x)$ from (5.2) or (5.3) by observing an actual $Q(t)$ or by doing a simulation, then to obtain an accuracy of about 10% with $\rho = 0.9$, one must observe $Q(t)$ for a time of order $100 T_0 \simeq 10^4$ $(1/\mu)$; for an accuracy of 1% one must observe it for about $10^4 T_0 \simeq 10^6 (1/\mu)$. It is usually rather absurd to try to estimate $F_Q(x)$ for $\rho = 0.99$ by simulation. An accuracy of 10% would require an observation time of about $10^6 (1/\mu)$; an accuracy of 1% would require about 10^8 $(1/\mu)$.

In typical applications the arrival process is not exactly stationary. The importance of the time scale T_0 relates more to the questions of how rapidly $\rho(t)$ can change and still have the queue behave as if it were (nearly) stationary with the parameter ρ evaluated as the $\rho(t)$ at time t.

Suppose that at time t the queue distribution was close to the equilibrium distribution associated with the value ρ at that time. If at some later time the value of ρ is slightly different, the system will try to adjust to the equilibrium distribution for the new value of ρ. If we think of the queue distribution at time t as a nonequilibrium initial queue distribution for the ρ at the later time, it will take a time of order T_0 for the queue distribution to adjust to the new ρ. A necessary condition for the queue distribution to keep up with the changing equilibrium is that the scale of queue length $L_0(\rho(t))$ at time t is nearly equal to the scale $L_0(\rho(t + T_0))$ at time $t + T_0$. 'Nearly equal' must mean in a relative sense, namely that

$$\left| \frac{L_0(\rho(t + T_0)) - L_0(\rho(t))}{L_0(\rho(t))} \right| \ll 1. \qquad (5.7)$$

If $\rho(t)$ is increasing and approaches 1, the time T_0 will increase rapidly. If, for example, $\rho(t)$ increases nearly linearly with time, so that $d\rho(t)/dt$ is nearly constant over a time of order T_0, the condition (5.7) requires that

$$\left|\frac{T_0}{L_0}\frac{dL_0(\rho(t))}{dt}\right| \simeq \frac{(1/\mu)|d\rho(t)/dt|}{[1-\rho(t)]^3} \ll 1, \qquad (5.7a)$$

i.e., the change in ρ during a time $(1/\mu)$ is small compared with $[1-\rho(t)]^3$.

Although the scale L_0 of the equilibrium queue distribution becomes infinite for $\rho \to 1$, the time required to reach the equilibrium also becomes infinite (at an even faster rate). For any nonzero (constant) rate of growth for $\rho(t)$, there will be a time when the equilibrium queue distribution is changing so rapidly that the actual queue cannot keep up.

There is, of course, nothing that prevents ρ from exceeding 1. Indeed, once ρ does exceed 1 the future evolution of $Q(t)$ is relatively simple. It grows at an average rate of $\lambda(t) - \mu$, but one must also know how large the queue is before it starts this growth. This problem will be discussed in Chapter 9.

For typical real processes $\rho(t)$ is likely to change appreciably over a time period of an hour or so, thus $|d\rho(t)/dt|$ is likely to be of the order of 10^{-2} per minute. Typical services times, however, cover a wide range. For many services such as bank tellers, airport runways, or telephone calls, $1/\mu$ is comparable with a minute. Thus $(1/\mu)|d\rho(t)/dt|$ may be of order 10^{-2} (which, in some sense, might be considered as 'small'). The condition (5.7a) would certainly fail if $[1-\rho(t)]^3 < (1/\mu)|d\rho(t)/dt| \simeq 10^{-2}$, but $\rho(t)$ need not be very close to 1 for the third power of $1-\rho(t)$ to be less than 10^{-2}. It is true already for $\rho(t) \gtrsim 0.8$. There are, however, some systems for which $1/\mu$ is measured in seconds (cars passing a point on a major highway, or certain tasks done by a computer) and for which $(1/\mu)|d\rho(t)/dt|$ may be of the order 10^{-4}. Even then, $[1-\rho(t)]^3$ is less than 10^{-4} for $\rho \gtrsim 0.95$. One must be quite cautious in using equilibrium queue distributions for applications in which ρ is close to 1.

5.3   Random walk

The study of equilibrium queue distributions is obviously an important part of the analysis of queues, but not as important as one might infer from the volume of literature devoted to it. The properties of queues which never vanish are relatively easy to analyze, and so are certain properties of equilibrium distributions. The mathematical methods used in these two extreme situations, however, are quite different, and the intermixing of the techniques to handle general time-dependent queues is not easy. The evaluation of equilibrium distributions is usually accomplished by means of mathematical techniques which give little insight into what is physically happening, the queue dynamics.

No attempt will be made here to survey the literature on equilibrium queue distributions. There already exist books at all levels of difficulty including many which survey nearly all of the solved problems (few emphasize, however, which problems have not been solved and why). We consider only the most elementary examples mostly to illustrate how equilibrium distributions can be evaluated while avoiding analysis of queue dynamics.

The first queueing model is rather artificial. In the probability literature it is more commonly described as a random walk, gambling model, or various other things, than as a model of a queue.

Suppose, as in problem 1.1, that events can occur only at integer times $t = 0, \pm 1, \pm 2, \ldots$ in some suitable time units (or at least we observe the system only at integer times). At time t (or between times $t - 1$ and t) there is at most one arrival which occurs with probability $\lambda, 0 < \lambda < 1$; no arrival occurs with probability $1 - \lambda$. The expected number of arrivals per unit time (the arrival rate) is λ.

The departure process is such that if a customer is in service or enters service at time t, he will leave at time $t + 1$ with probability μ, and will remain in service with probability $(1 - \mu)$. If there is a nonzero queue, a customer who leaves the service at time $t + 1$ is immediately replaced by a new customer. The server can serve only one customer at a time, so the expected number of departures per unit time (while the service is operating) is μ. We assume that $\lambda < \mu$.

It is convenient here, as in most queueing models, to deal with the number of customers in the system $Q_s(t)$ rather than the number in the queue $Q(t)$. The state $Q_s(t) = 0$ represents no queue and no customers in service; otherwise $Q_s(t) = Q(t) + 1$.

Between times t and $t + 1$, $Q_s(t)$ can increase by 1 if there is an arrival but no departure; remain the same if either there is no arrival and no departure or one arrival and one departure; or decrease by 1 if there is no arrival and one departure. If $Q_s(t) \geq 1$

$$Q_s(t + 1) = Q_s(t) + X(t), \tag{5.8}$$

with $X(t)$ (the change in queue) having the distribution

$$X(t) = \begin{cases} 1 & \text{with probability} & \lambda(1-\mu) \\ 0 & \text{with probability} & (1-\mu)(1-\lambda)+\lambda\mu \\ -1 & \text{with probability} & (1-\lambda)\mu. \end{cases} \quad (5.9)$$

If, however, $Q_s(t) = 0$, there can be no departures and

$$Q_s(t+1) = \begin{cases} 1 & \text{with probability} & \lambda \\ 0 & \text{with probability} & 1-\lambda. \end{cases} \quad (5.10)$$

If we knew that the queues were sufficiently large that there was negligible probability for the queue to vanish during some period of time, then we could disregard the 'boundary condition' (5.10) and iterate (5.8).

$$\begin{aligned} Q_s(t+1) &= Q_s(t) + X(t) \\ &= Q_s(t-1) + X(t-1) + X(t) \\ &= Q_s(t-2) + X(t-2) + X(t-1) + X(t) \end{aligned}$$

etc.

Since the $X(t)$ are independent random variables, the sum of sufficiently many of the $X(t)$ will be approximately normal. Thus, we can reconfirm the behavior described in Chapter 4.

To treat the problem with the boundary condition, let

$$p_j(t) = P\{Q_s(t) = j\}$$

be the marginal distribution of $Q_s(t)$. The distribution of $Q_s(t+1)$ can be related to that of $Q_s(t)$ through (5.8)–(5.10):

$$\begin{aligned} p_j(t+1) &= \lambda(1-\mu)p_{j-1}(t) + [(1-\mu)(1-\lambda)+\lambda\mu]p_j(t) \\ &\quad + (1-\lambda)\mu p_{j+1}(t) \quad \text{for } j \geq 2 \\ p_1(t+1) &= \lambda p_0(t) + [(1-\mu)(1-\lambda)+\lambda\mu]p_1(t) \\ &\quad + (1-\lambda)\mu p_2(t) \\ p_0(t+1) &= (1-\lambda)p_0(t) + (1-\lambda)\mu p_1(t). \end{aligned} \quad (5.11)$$

The first equation says that the probability of having $Q_s(t+1) = j$ is the probability that at time $t, Q_s(t) = j-1$ and $X(t) = +1$, plus the probability that at time $t, Q_s(t) = j$ and $X(t) = 0$, etc.

For any given values of λ and μ and any initial distributions $p_k(0)$, $k = 0, 1, 2, \ldots$, one can evaluate $p_j(1), j = 0, 1, 2, \ldots$, from (5.11); then from $p_j(1)$, evaluate $p_j(2)$, etc. By a straightforward iteration one can evaluate $p_j(t)$ for all finite j and t. In principle, there is no difficulty in evaluating the evolution of the distribution for $Q_s(t)$ *exactly*. But, as is frequently the case in applied mathematics, the difficulty lies in the *approximate* solution and in the interpretation of the qualitative behavior; how it depends upon the parameters λ and μ, and what of practical value one is to learn from it.

The neatness of (5.11) is deceptive. The behavior of the $p_j(t)$ is actually quite complicated. Somehow this equation must describe all those properties which we have previously claimed that queues should have. In appropriate situations it must show the properties of fluid approximations described in Chapter 2, and some of the normal distributions described in Chapter 4. If $\lambda < \mu$, the distribution should also approach an equilibrium distribution such that for $t \to \infty$, $p_j(t) \to p_j$. Equation (5.11) must not only describe this equilibrium distribution, but also how it evolves.

We can by-pass the question of how the equilibrium distribution is attained, however, and simply ask: What distribution p_j has the property that if $p_j(t) = p_j$, so is $p_j(t+1) = p_j$? If in (5.11) we replace $p_j(t)$ by p_j in both sides of the equation, we obtain an infinite system of simultaneous linear equations for the $p_j, j = 0, 1, \ldots$. Because of the special structure of this system of equations, the p_j can be evaluated iteratively.

If we rearrange the terms of (5.11) into the form

$$\mu(1-\lambda)p_{j+1} - \lambda(1-\mu)p_j = \mu(1-\lambda)p_j - \lambda(1-\mu)p_{j-1}, \quad \text{for } j \geq 2$$

$$\mu(1-\lambda)p_2 - \lambda(1-\mu)p_1 = \mu(1-\lambda)p_1 - \lambda p_0$$

$$\mu(1-\lambda)p_1 - \lambda p_0 \qquad = 0,$$

the right hand side of the second equation vanishes by virtue of the third equation. The right hand side of the first equation therefore vanishes for $j = 2$ and for all $j > 2$.

From this we can evaluate all the p_j in terms of p_0;

$$p_j = p_0 \frac{\lambda}{(1-\lambda)\mu} \left[\frac{\lambda(1-\mu)}{\mu(1-\lambda)} \right]^{j-1} \quad j \geq 1. \tag{5.12}$$

Finally, we can determine p_0 from the condition that $\sum_0^\infty p_j = 1$:

$$1 = p_0 + \sum_{j=1}^\infty p_0 \frac{\lambda}{(1-\lambda)\mu} \left[\frac{\lambda(1-\mu)}{\mu(1-\lambda)} \right]^{j-1}$$

$$= p_0 \left[1 + \frac{\lambda}{(1-\lambda)\mu \left(1 - \frac{\lambda(1-\mu)}{\mu(1-\lambda)} \right)} \right]$$

$$= p_0 \frac{\mu}{\mu - \lambda},$$

i.e,

$$p_0 = 1 - \lambda/\mu = 1 - \rho. \tag{5.13}$$

This formula, of course, applies for any single-channel server; the server must be idle a fraction $1 - \rho$ of the time.

Except for the state $j = 0$, the distribution of Q_s is geometric, with

$$E\{Q_s\} = \frac{\lambda(1-\lambda)}{\mu - \lambda} = \rho\frac{(1-\lambda)}{(1-\rho)}. \tag{5.14}$$

We have not shown here how long it takes to reach the equilibrium but (5.14) does confirm that $E\{Q_s\}$ and $E\{Q\} \to \infty$ for $\rho \to 1$ like $(1-\rho)^{-1}$. For these arrival and service processes

$$I_D = 1 - \mu \quad \text{and} \quad I_A = 1 - \lambda$$

and the L_0 of (5.5) has the value

$$L_0 = \frac{\mu(1-\mu) + \lambda(1-\lambda)}{\mu(1-\rho)}.$$

This length scale is meaningful only for ρ close to 1, i.e., $\mu \simeq \lambda$ for which

$$L_0 \simeq \frac{2(1-\mu)}{1-\rho} \simeq 2E\{Q_s\}. \tag{5.15}$$

5.4 The M/M/1 queue

A queueing system in continuous time very closely related to the above discrete time queue is one known as the M/M/1 queue. The first M signifies, in the usual classification of queues, that the time intervals between arrivals are independent and exponentially distributed (memoryless). The arrival process is a homogeneous Poisson process with rate λ. The second M signifies that the service times are also independent random variables with an exponential distribution (service rate μ). The 1 signifies that there is only one channel, i.e., customers are served one at a time.

The most important property of exponentially distributed random variables (for example, the service time S in the present system) and the property which makes the analysis of the M/M/1 queue relatively simple, is the following. If S is a random variable with parameter μ and one observes that the service has not been completed by a time t after it started, the remaining service time after time t is also exponentially distributed with parameter μ, independent of t. Thus, if one observes that a customer is being served, his past history (when he entered service) is irrelevant to the prediction of the future (when the service will end).

This follows from the fact that

$$P\{S-t > y | S > t\} = \frac{P\{S-t > y\}}{P\{S > t\}} = \frac{e^{-\mu(y+t)}}{e^{-\mu t}} = e^{-\mu y}.$$

It also follows that there is a probability $\mu \, dy$ that a service in operation at time t will be completed between time t and $t+dy$, independently of t.

To exploit this property we notice that if $Q_s(t) = j \geq 1$ at time t, then there is a probability $\lambda \, dt$ that between time t and $t+dt$ a customer will arrive and increase the queue to $j+1$; there is a probability $\mu \, dt$ that the customer in service will leave and reduce the queue to $j-1$; and there is a probability $1 - (\lambda + \mu) \, dt$ that the queue will not change. Thus, if $p_j(t)$ again represents the probability that $Q_s(t) = j$, then for $j \geq 1$

$$p_j(t+dt) = [1 - (\lambda + \mu) \, dt] p_j(t) + \lambda \, dt \, p_{j-1}(t) + \mu \, dt \, p_{j+1}(t)$$

or

$$\frac{d p_j(t)}{dt} = -(\lambda + \mu) p_j(t) + \lambda p_{j-1}(t) + \mu p_{j+1}(t). \qquad (5.16)$$

Similarly,

$$\frac{d p_0(t)}{dt} = -\lambda p_0(t) + \mu p_1(t). \qquad (5.16a)$$

This is a system of ordinary differential equations. If there exists a stationary distribution, it must be a solution of (5.16), (5.16a) with $p_j(t) = p_j$. Again we see, as with (5.11), that we can iteratively evaluate $p_1, p_2, \ldots$ in terms of p_0. Equation (5.16a) gives

$$p_1 = (\lambda/\mu) p_0 = \rho \, p_0;$$

substitution of this in (5.15) for $j = 1$ gives $p_2 = (\lambda/\mu) p_1$, etc. Generally

$$p_j = \rho \, p_{j-1} = \rho^j p_0 = (1-\rho) \rho^j \qquad (5.17)$$

a distribution which is (completely) geometric with

$$E\{Q_s\} = \rho/(1-\rho). \qquad (5.18)$$

This differs from (5.14) only by the factor $1 - \lambda$, but for the M/M/1 system $I_A = I_D = 1$, so that again $L_0 \simeq 2E\{Q_s\}$. It is also interesting to observe that in (5.14), $E\{Q_s\}$ depends upon both λ and ρ, but (5.17) depends only upon ρ. For the random walk model there is a time scale (the integers) relative to which service times can be measured, but for the M/M/1 system there is no natural unit² of time. Since the latter queue distribution must be independent of what units are used for measuring λ and μ, it can depend only upon the dimensionless ratio ρ.

5.5 The M/M/m queue

If we are concerned with queueing systems which typically contain a large number of customers, the behavior of an m-channel service with only a few channels, each of service rate μ, should not differ very much from a hypothetical single-channel server of service rate $m\mu$. One inescapable difference, of course, is that a multiple-channel server is capable of storing several customers at a time in the service. If the server has a large number of channels, the server itself may store a significant fraction of all customers in the system. We saw in Section 3.7 that one might even apply fluid approximations to the customers within the service.

To illustrate some of the differences between single- and multiple-channel systems, we consider here the simplest (mathematically) type multiple-channel system, M/M/m. The arrival process is a homogeneous Poisson process of rate λ, and there are m servers, each with an exponentially distributed service time of rate μ, statistically independent of each other.

If there are j customers in service at time t, each with an exponential distribution of service time, then the time of completion of any service is statistically independent of any past history. There is a probability $\exp(-\mu y)$ that the remaining service time S_k of the kth service will last at least a time y. The probability that none of the j customers completes service within a time y is

$$P\{S_1 > y, S_2 > y, \ldots, \text{and } S_j > y\}$$
$$= P\{S_1 > y\}P\{S_2 > y\} \ldots P\{S_j > y\} = \exp(-j\mu y).$$

This is also the probability that the next service occurs after time $t + y$, i.e.,

$$P\{\text{time until the next completed service} > y\} = e^{-j\mu y}.$$

The j servers that are in service at time t thus act like a single server serving at a rate $j\mu$, and there is a probability $j\mu\,dy$ that a service will be completed between time t and $t+dy$. The rate of service depends upon the 'state of the system', how many customers, j, are in service.

If the number of customers in queue and in service $Q_s(t) = j$, then there is a probability $\lambda\,dt$ that a customer will arrive within a time dt and increase $Q_s(t)$ to $j+1$. For $j \le m$ there is a probability $j\mu\,dt$ that a customer will leave the service and reduce $Q_s(t)$ to $j-1$. For $j \ge m$, this happens with probability $m\mu\,dt$. If, as in Section 5.4, we let $p_j(t)$ denote the probability that $Q_s(t) = j$, then the generalization of (5.16) will yield the equations:

$$dp_j(t)/dt = -(\lambda + \mu_j)\,p_j(t) + \lambda\,p_{j-1}(t) + \mu_{j+1}\,p_{j+1}(t)$$
$$j = 0, 1, \ldots \tag{5.19}$$

with

$$\mu_j = \begin{cases} j\mu & j \le m \\ m\mu & j \ge m \end{cases} \quad \text{and} \quad p_{-1}(t) \equiv 0.$$

The stationary distribution for this system of differential equations is obtained, as before, by setting the derivatives equal to zero and iterating the resulting linear equations for the $p_j(t) = p_j$. The p_j will have the form

$$p_j = \begin{cases} \dfrac{(m\rho)^j}{j!}\,p_0 & \text{for } j \le m \\[2ex] \rho^{j-m}\,p_m = \rho^{j-m}(m\rho)^m\,p_0/m! & \text{for } j \ge m \end{cases} \tag{5.20}$$

with $\rho \equiv \lambda/(m\mu)$. The p_0 is to be determined so that $\sum_j p_j = 1$.

In the special case $m = 1$, (5.20) simplifies to (5.17), whereas for $m \to \infty$

$$\sum_{j=0}^{m} p_j \to \sum_{j=0}^{\infty} \frac{1}{j!}\left(\frac{\lambda}{\mu}\right)^j p_0 = p_0\,e^{\lambda/\mu} = 1$$

and

$$p_j = \frac{1}{j!}\left(\frac{\lambda}{\mu}\right)^j e^{-\lambda/\mu}, \quad m = \infty \tag{5.21}$$

a Poisson distribution with parameter λ/μ. For $1 < m < \infty$, successive values of p_j start as if they were going to generate a Poisson distribution, until $j = m$. For $j > m$, the p_j suddenly switch to the form of the geometric distribution with parameter ρ characteristic of a single-channel server with service rate $m\mu$.

For $j \leq m$, the p_j satisfy the relation

$$\frac{p_{j+1}}{p_j} = \frac{m\rho}{(j+1)}, \quad j \leq m.$$

Thus the p_j are increasing with j, $p_{j+1} > p_j$ for $j+1 < m\rho$ but decreasing for $j+1 > m\rho$. For an equilibrium distribution to exist, it is necessary that $\rho < 1$, therefore $p_m/p_{m-1} = \rho < 1$. Also, $p_{j+1}/p_j = \rho$ for $j \geq m$.

For $j > m$ the p_j do not decrease as rapidly as for the Poisson distribution. It follows that, for given values of λ and μ, the random equilibrium number of customers in the system, Q_s, is monotone decreasing with m in the stochastic sense that $P\{Q_s > j\}$ is monotone decreasing in m for every j; the more servers, the less the delay.

If we compare (5.20) with the geometric distribution for a hypothetical single-channel server with service rate $m\mu$ we see that the geometric distribution has more probability in the range $j \leq m$. The number of customers in the system Q_s for the $M/M/m$ queue is, therefore, stochastically larger than for the single-channel server of service rate $m\mu$, but the number in the queue Q is less.

We know that for any m larger than λ/μ, a server can handle all arrivals with finite delay but in designing a facility, particularly with $\lambda/\mu \gg 1$, we might also like to know how much larger than λ/μ, m must be in order to avoid excessive delays.

Since a Poisson distribution with parameter $\lambda/\mu \gg 1$ can be approximated by a normal distribution with mean λ/μ and variance λ/μ, it follows from (5.20) that

$$p_j \simeq A \exp\left[\frac{-(j-\lambda/\mu)^2}{2(\lambda/\mu)}\right] \quad \text{for } j \leq m \tag{5.22}$$

for some constant A, and the geometric distribution for $j \geq m$ matches the distribution both in amplitude and slope at $j = m$. If

$$(m - \lambda/\mu)/(\lambda/\mu)^{1/2} \gtrsim 2,$$

then the range $j \leq m$ extends over the peak of the normal distribution and far into its tail. Any probability in the geometric distribution for $j \geq m$ is not only of small amplitude but also decays rapidly. It suffices, therefore, to choose an m equal to λ/μ plus one or two times the standard deviation $(\lambda/\mu)^{1/2}$ in order to guarantee a small probability for any queueing delays.

Even for

$$m \gtrsim \lambda/\mu + (\lambda/\mu)^{1/2},$$

the geometric continuation of the distribution for $j \geq m$ remains quite close to the continuation of the normal distribution beyond $j = m$, so that (5.22) is a reasonable approximation for all j. Also

$$\sum_j p_j \simeq A \int_{-\infty}^{+\infty} dj \exp\left[-\frac{(j-\lambda/\mu)^2}{2(\lambda/\mu)} \right] = A\,(2\pi\lambda/\mu)^{1/2} = 1,$$

so

$$p_j \simeq \frac{1}{(2\pi\lambda/\mu)^{1/2}} \exp\left[\frac{-(j-\lambda/\mu)^2}{2(\lambda/\mu)} \right]. \tag{5.23}$$

The amount of queueing can be inferred from the tail of this distribution with $j \geq m$.

If

$$m - \lambda/\mu < (\lambda/\mu)^{1/2},$$

in particular if $m - \lambda/\mu$ is small compared with $(\lambda/\mu)^{1/2}$, (5.23) is no longer valid. The geometric distribution starts with a slow rate of decay and (unlike the normal distribution) keeps it. Most of the probability is in the geometric part of the distribution and

$$p_j \simeq \rho^{j-m} p_m$$
$$\sum_{j=0}^{\infty} p_j \simeq \sum_{j=m}^{\infty} p_j = p_m (1-\rho)^{-1} = 1$$
$$p_j \simeq (1-\rho)\rho^{j-m} \quad j \geq m \tag{5.24}$$

with a negligible total probability for all $j < m$. The system behaves essentially as a single-server system with service rate $m\mu$.

Actually, (5.24) is a rather crude approximation; it neglects completely the probability for states $j < m$. A better approximation results if one joins a geometric (exponential) distribution smoothly to (5.22) and then rescales this distribution by dividing it by $\Sigma_j p_j$. This sum can be approximated by an integral or evaluated numerically.

Fig. 5.1 illustrates how the queue distribution changes with m for a fixed average number of busy servers $\lambda/\mu = 16$. Discrete values of p_j versus j are joined by line segments to approximate a continuous distribution. The solid lines represent the exact distributions for $m = \infty$, 22, 18, and 17; the broken lines are obtained by joining a normal distribution (5.23) smoothly to an exponential.

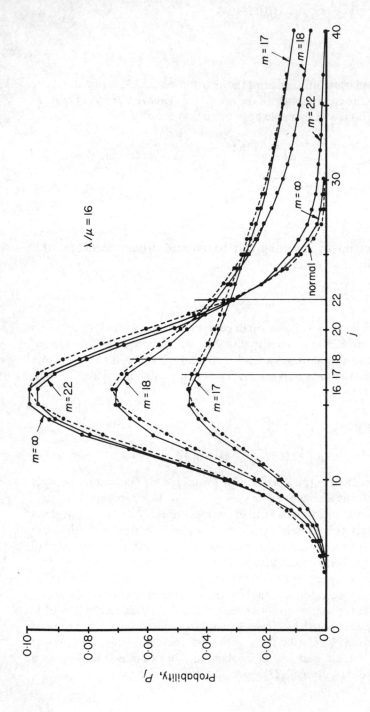

Figure 5.1 Points joined by solid lines represent the probability p_j for j customers in an m-channel server with $\lambda/\mu = 16$. Points joined by a broken line represent approximate values obtained by joining a normal distribution to an exponential distribution.

The curve for

$$m = 22 = \lambda/\mu + (3/2)\,(\lambda/\mu)^{1/2}$$

shows that the p_j stay close to those for $m = \infty$ even for several j values beyond m. As m decreases to 18 and 17, the distribution for $j \leq 16$ retains the same shape but is scaled down as most of the probability goes into the goemetric distribution for $j > m$.

5.6   The M/M/m/c system

As a simple variation on the results of the previous section, suppose that there is storage space for only c customers including those in service. If a customer arrives and finds the system full, he goes away. If $c \geq m$ there is room for $c - m$ customers in the queue. For $c < m$ the system behaves as if there were only c servers and no room for a queue; it suffices therefore to assume that $c \geq m$.

For Poisson arrivals and exponentially distributed service times (5.19) is still valid for $j \leq c - 1$ but the equation for $p_c(t)$ becomes

$$\frac{dp_c(t)}{dt} = -\mu_c p_c(t) + \lambda p_{c-1}(t) = -m\mu p_c(t) + \lambda p_{c-1}(t).$$

If the system is in state c and someone arrives, nothing changes.

The equilibrium solution (5.20) is still valid for $j \leq c$. The only difference is that the normalization condition is now

$$\sum_{j=0}^{c} p_j = 1.$$

Also, the solution (5.20) is no longer restricted by a condition that $\rho = \lambda/m\mu$ be less than 1. If $\rho > 1$, $p_{j+1}/p_j > 1$ for all j; most of the probability lies near the state $j = c$.

For the single-channel server, the queue distribution is a truncated geometric distribution

$$p_j = \frac{(1-\rho)\rho^j}{1-\rho^{c+1}}, \quad 0 \leq j \leq c. \tag{5.25}$$

For $\rho > 1$, however, it might be more convenient to consider the empty storage spaces than the occupied ones. We can also write (5.25) in the form

$$p_{c-k} = \frac{[1-(1/\rho)]\,(1/\rho)^k}{[1-(1/\rho)^{c+1}]}.$$

which has exactly the same form as (5.25) except that ρ is replaced by $1/\rho$, i.e., μ and λ are interchanged.

This is another illustration of the symmetry discussed in Section 3.5. Imagine that the arrival process were generated by a zeroth server serving customers from an infinite reservoir at rate λ and exponentially distributed service time, but the zeroth server cannot operate if the storage is full. The storage also feeds a server numbered one which operates at rate μ provided that the storage is not empty. If we look at the flow of vacancies, this system is obviously equivalent to one in which vacancies travel in the opposite direction, through server one into the storage and then through server zero. They can pass server one at rate μ provided that the storage is not empty (full of vacancies) and can leave through server zero at rate λ provided that the storage is not full (empty of vacancies). From the point of view of the vacancies, the system behaves as if objects (vacancies) arrived at a rate μ and were served at a rate λ but there was storage space for only c vacancies.

Regardless of whether we consider this as a model for a system in which customers arrive at a rate λ and some are turned away or there really is a zeroth server serving an infinite reservoir, one is likely to be most interested in the equilibrium rate at which customers actually pass through the system or the rate at which they are turned away. Of course, the equilibrium rate at which they pass server zero (enter the system) must be the same as the rate at which they leave. Whenever the storage is full, the service rate of server zero vanishes and whenever the storage is empty, the service rate of server one vanishes. For the single-channel server

$$\frac{dE\{A(t)\}}{dt} = \frac{dE\{D(t)\}}{dt} = \lambda(1 - p_c) = \mu(1 - p_0)$$

$$\begin{aligned} &= \lambda(1 - \rho^c)/(1 - \rho^{c+1}) \\ &= \mu[1 - (1/\rho)^c]/[1 - (1/\rho)^{c+1}] \quad \text{for} \quad \rho \neq 1 \\ &= \mu c/(c+1) \quad\quad\quad\quad\quad\quad \text{if} \quad \rho = 1, \mu = \lambda. \end{aligned}$$

$$\text{(5.26)}$$

This is of particular interest as applied to situations in which there really are two servers in series with a finite storage between them because (5.26) represents the maximum average rate at which customers can pass through the system. This rate is generally less than the service rate of either server for $c < \infty$. Particularly for $\lambda = \mu$, the loss in service rate due to blocking, $\mu/(c+1)$, does not decrease very rapidly with c. In designing a system with two servers in series, one of which is very expensive, it would be advantageous, if possible, to provide a larger service rate for the cheap server and enough storage space between the servers that the expensive server is kept busy.

For $0 < 1 - \rho \ll 1$ the loss in service rate due to blocking

$$\lambda - \frac{dE\{A(t)\}}{dt} = \frac{\lambda \rho^c (1 - \rho)}{1 - \rho^{c+1}} \simeq \frac{\lambda e^{-c(1 - \rho)}(1 - \rho)}{1 - e^{-(c+1)(1 - \rho)}}$$

This will decrease rapidly with c for $c(1 - \rho) \gtrsim 1$. Since the mean queue length with $c = \infty$ would be approximately $(1 - \rho)^{-1}$, it is obvious that the storage space c necessary to prevent most blocking should be of this order of magnitude.

For the multiple-channel server it is still true that

$$\lambda - \frac{dE\{A(t)\}}{dt} = \lambda p_c$$

but the formula for p_c is more complicated because of the normalization. It is also true that

$$\frac{dE\{A(t)\}}{dt} = \frac{dE\{D(t)\}}{dt} = \mu \sum_{j=0}^{m} jp_j + \mu \sum_{j=m+1}^{c} mp_j.$$

5.7   M/G/1 system

The customary methods used to analyze systems with Poisson arrivals and exponentially distributed service times exploits the fact that, at any time t, the future behavior of the system is determined if one specifies only the count of customers in the queue and in the service, $Q_s(t)$. How long a customer has been in service or when the customers arrived prior to time t is irrelevent.

The most common general methods for analyzing equilibrium queues makes use of the theory of Markov processes. One first tries to identify a 'state space'. A state of the system at some time t would consist of a specification of any properties of the system from which one could uniquely infer the probability distribution of the state at any (or some) future time. If Z_t is the state of the system at time t, a (random) point in the state space, then there exist some conditional probabilities $P\{Z_{t+\tau} | Z_t\}$ having the property that for any sequence of times $0 < \tau_1 < \tau_2 < \ldots < \tau_n$ the joint probability distribution of $Z_{t+\tau_1}, Z_{t+\tau_2}, \ldots$ can be written in the form

$$\begin{aligned}
P\{Z_t, & Z_{t+\tau_1}, Z_{t+\tau_2}, \ldots, Z_{t+\tau_n}\} \\
&= P\{Z_t\} P\{Z_{t+\tau_1} | Z_t\} P\{Z_{t+\tau_2} | Z_{t+\tau_1}\} \times \cdots \\
&\quad \times P\{Z_{t+\tau_n} | Z_{t+\tau_{n-1}}\}
\end{aligned} \tag{5.27}$$

with each factor depending upon (at most) the state at two times. In particular, the probability of being in state $Z_{t+\tau}$ at time $t+\tau$ is determined from an equation of the form

$$P\{Z_{t+\tau}\} = \int P\{Z_{t+\tau}|Z_t\} \, dP\{Z_t\} \qquad (5.28)$$

i.e., the distribution at time $t+\tau$ is obtained by averaging the conditional distribution over all states Z_t at time t. Also, the joint distribution of $Z_{t+\tau_1}, Z_{t+\tau_2}, \ldots$ can be determined by integrating (5.27) over $P\{Z_t\}$.

If the Markov process described above has stationary transition probabilities, $P\{Z_{t+\tau}|Z_t\}$ is independent of t. A stationary distribution would be a distribution $P\{Z_t\}$ having the property that $P\{Z_{t+\tau}\} = P\{Z_t\}$ for all $\tau > 0$. For most of the processes of interest here we expect that the distribution $P\{Z_{t+\tau}\}$ will approach an equilibrium distribution independent of $P\{Z_t\}$ for $\tau \to \infty$, and that the equilibrium distribution can be evaluated from (5.28).

This is essentially the type of argument used above for the $M/M/m$ system. The state space was the space of integers; the state of the system was the number of customers in the system; and the probability distribution of $Q_s(t)$ was denoted by $p_j(t)$. Although we did not determine explicitly the conditional probabilities of going from state j at time t to k at time $t+\tau$ for all τ, we did evaluate the transitions in an infinitesimal time, i.e., the transition rates. The analogue of (5.28) was a differential equation for $dp_j(t)/dt$. The equilibrium distribution was obtained by solving the system of equations with $dp_j(t)/dt = 0$, i.e., $p_j(t+\tau) = p_j(t)$.

For many (perhaps most) types of queueing systems, even those involving several customer types and/or servers, it is possible to identify a finite dimensional state space, i.e., a random state vector Z_t which (at least approximately) satisfies (5.27). The class of problems for which one can evaluate the equilibrium distribution of Z_t is, however, quite limited. There may, of course, be some question as to whether an equilibrium distribution exists, but the goal in most queueing problems is to identify some state variables (which may actually give only a partial description of the system behavior) for which equation (5.28), or its equivalent, can be analyzed in some detail. In most cases one tries to identify a one-dimensional discrete or continuous state space since very few multidimensional problems are tractable.

For an $M/G/1$ system (Poisson arrivals and general independent service times), for example, it is obvious that a specification of the number of customers in the system at time t and the length of time the customer in service (if any) has been in service is sufficient to determine the future stochastic properties of these variables. This appears to be a two-dimensional problem with one integer and one continuous state variable, but the latter variable has very special properties. During a time t to $t + dt$ it can either increase by dt or go to zero. If we look at the system only at those times when a service is completed, it suffices to identify only the number of customers in the system at that time.

Let

$Q_s(t_{sn})$ = number of customers in the system immediately after n customers have completed service.

If $Q_s(t_{sn}) \geq 1$, the next service completion time is

$$t_{s, n+1} = t_{sn} + S_{n+1} \tag{5.29}$$

with S_{n+1} the service time of the $n + 1$ customer to leave, and S_{n+1} is independent of t_{sn} (or any previous events). Also

$$Q_s(t_{s, n+1}) = Q_s(t_{sn}) - 1 + [A(t_{s, n+1}) - A(t_{sn})]. \tag{5.30}$$

Since the arrivals define a Poisson process, $[A(t_{s, n+1}) - A(t_{sn})]$ is independent of $Q_s(t_{sn})$ or t_{sn}. For given S_{n+1} it has a Poisson distribution

$$P\{A(t_{s, n+1}) - A(t_{sn}) = k \,|\, S_{n+1}\} = \frac{(\lambda S_{n+1})^k}{k!} \exp(-\lambda S_{n+1})$$

and so

$$P\{A(t_{s, n+1}) - A(t_{sn}) = k)\} = \int_0^\infty \frac{(\lambda s)^k}{k!} e^{-\lambda s} dF_s(s). \tag{5.31}$$

If, however, $Q_s(t_{sn}) = 0$ the next service completion is at time

$$t_{s, n+1} = t_{n+1} + S_{n+1} = t_{sn} + (t_{n+1} - t_{sn}) + S_{n+1}$$

in which t_{n+1} is the arrival time of the $n + 1$th customer. Since the arrivals are a Poisson process, the time $t_{n+1} - t_{sn}$ until the next arrival is exponentially distributed with rate λ independent of any previous history. Also for $Q_s(t_{sn}) = 0$,

$$Q_s(t_{s, n+1}) = A(t_{s, n+1}) - A(t_{n+1}) \tag{5.32}$$

is distributed as (5.31), independent of any past history.

Given $Q_s(t_{sn})$ and t_{sn}, the above equations will determine the joint distribution of $Q_s(t_{s,\,n+1})$ and $t_{s,\,n+1}$, but also the value of $Q_s(t_{sn})$ alone will determine the distribution of $Q_s(t_{s,\,n+1})$. The sequence $Q_s(t_{sn})$ is commonly described as an 'embedded' Markov chain.

The right hand sides of (5.30) and (5.32) describe $Q_s(t_{s,\,n+1})$ almost like a sum of independent random variables, except that the state $Q_s(t_{sn}) = 0$ must be treated separately. The distribution of the sum of two independent random variables is, generally, rather awkward to manipulate (it is the convolution of the distributions for the two variables), but the probability generating function of the sum of two independent variables is simply the product of two generating functions. It seems natural, therefore, to try to evaluate the generating function of $Q_s(t_{s,\,n+1})$ in terms of the generating function of $Q_s(t_{sn})$. If

$$p_j(t_{sn}) = P\{Q_s(t_{sn}) = j\}$$

then the probability generating function of $Q_s(t_{sn})$ is defined as

$$G_n(z) = \sum_{j=0}^{\infty} p_j(t_{sn})z^j = E\{z^{Q_s(t_{sn})}\}.$$

The generating function for $Q_s(t_{s,\,n+1})$ can be written as

$$G_{n+1}(z) = \sum_{j=0}^{\infty} p_j(t_{sn})E\{z^{Q_s(t_{s,\,n+1})}|Q_s(t_n) = j\}.$$

From (5.30) and (5.32),

$$G_{n+1}(z) = p_0(t_{sn})E\{z^{A(t_{s,\,n+1})-A(t_{n+1})}\}$$
$$+ \sum_{j=1}^{\infty} p_j(t_{sn})E\{z^{Q_s(t_{sn})-1+A(t_{s,\,n+1})-A(t_{sn})}|Q_s(t_n) = j\}.$$

Since the arrivals in time S_{n+1} are distributed as (5.31), independent of $Q_s(t_{sn})$, the expectation over the arrivals gives

$$G_{n+1}(z) = \left\{\sum_{k=0}^{\infty}\int_0^{\infty} \frac{(\lambda sz)^k}{k!}e^{-\lambda s}dF_S(s)\right\}\left[p_0(t_{sn})\right.$$
$$+ \sum_{j=1}^{\infty} p_j(t_{sn})z^{j-1}\right]$$
$$= \left\{\int_0^{\infty}\exp(-\lambda s(1-z))dF_S(s)\right\}[G_n(z)$$
$$- p_0(t_{sn})(1-z)]/z. \qquad (5.33)$$

The first factor of (5.33) can be considered a known function of z. If

$$\overline{F}_S(u) \equiv \int_0^\infty e^{-us} dF_S(s),$$

represents the Laplace transform of $F_S(s)$, then this factor is $\overline{F}_S(\lambda(1-z))$. A specification of $G_n(z)$ automatically gives $p_0(t_{sn}) = G_n(0)$, consequently (5.33) describes a rather simple relation between the generating function G_{n+1} of $Q_s(t_{s,\,n+1})$ and the generating function $G_n(z)$ of $Q_s(t_{sn})$.

Although (5.33) describes the transient behavior of $G_n(z)$ in the sense that one could iteratively evaluate $G_1(z), G_2(z), \ldots$, it is very difficult to see from (5.33) what is actually happening. To investigate transient behavior it would generally be simpler and more instructive to simulate the queue evolution from (5.30), (5.31) and (5.32). For $\rho = \lambda E\{S\} < 1$ we expect, however, that there will be an equilibrium distribution and consequently an equilibrium generating function for which

$$G_{n+1}(z) = G_n(z) = G(z)$$

with

$$G(z) = \overline{F}_S(\lambda(1-z))[G(z) - G(0)(1-z)]/z,$$

i.e.,

$$G(z) = G(0)(1-z) + \frac{G(0)(1-z)z}{\overline{F}_S(\lambda(1-z)) - z}. \qquad (5.34)$$

Although one can evaluate the equilibrium probabilities p_j by expanding G in powers of z,

$$G(z) = \sum_{j=0}^\infty p_j z^j, \qquad (5.35)$$

it is generally easier to evaluate the moments (or factorial moments) of Q_s from

$$\left.\frac{d^m G(z)}{dz^m}\right|_{z=1} = \sum_{j=0}^\infty j(j-1)\ldots(j-m+1)p_j$$

$$= E\{Q_s(Q_s-1)\ldots(Q_s-m+1)\}.$$

This is true mainly because the expansion of $\overline{F}_S(\lambda(1-z))$ around $z = 1$ has a simpler interpretation than the expansion around $z = 0$. The expansion around $z = 1$ gives

$$\overline{F}_S(\lambda(1-z)) = \int_0^\infty \exp(\lambda s(z-1)) dF_S(s) = \sum_{k=0}^\infty \frac{\lambda^k E\{S^k\}}{k!}(z-1)^k.$$

For $z \to 1$ both the numerator and denominator of the second term in (5.34) vanish, but one can also write.

$$\bar{F}_s(\lambda(1-z)) - z = (1-z)\left[1 - \sum_{k=1}^{\infty} \frac{\lambda^k E\{S^k\}}{k!}(z-1)^{k-1}\right].$$

Consequently,

$$G(1) = \frac{G(0)}{1-\rho}.$$

The normalization of the p_j, however, implies that $G(1) = 1$. Therefore

$$G(0) = 1 - \rho.$$

and

$$G(z) = (1-\rho)(1-z) + \frac{(1-\rho)z}{1 - \sum_{k=1}^{\infty} \lambda^k E\{S^k\}(z-1)^{k-1}/k!}.$$

(5.36)

Terms in the expansion of the right hand side of (5.36) in powers of $(z - 1)$ become successively more cumbersome but, generally, the kth term in the expansion of $G(z)$ involves moments of S only up to order $k + 1$. We conclude that the equilibrium kth moment of Q_s also depends upon moments of S up to order $k + 1$. In particular,

$$E\{Q_s\} = \rho + \frac{\lambda^2 E\{S^2\}}{2(1-\rho)} = \rho + \frac{\rho^2(1+C^2(S))}{2(1-\rho)}. \qquad (5.37)$$

Since the first term of this can be interpreted as the expected number of customers in the service, the expected number in the queue is

$$E\{Q\} = \frac{\rho^2(1+C^2(S))}{2(1-\rho)}.$$

There is no reason to have expected that there should be a simple relation between the equilibrium queue distribution p_j and a general service time distribution $F_s(s)$. Despite the fact that this solution may not be in a form that readily lends itself to numerical evaluation of the p_j, it does describe some important properties.

We know that $G(z)$ and all its derivatives are positive increasing functions of z for $z > 0$ and that (5.35) converges for $0 \le z \le 1$ with $G(1) = 1$. If $G(z)$ is bounded for $0 \le z < z_0$ for some $z_0 > 1$, it would be necessary that $p_j z^j \to 0$ for $j \to \infty$ for $z < z_0$, i.e., the p_j must go to zero at least as fast as z^{-j} (like a geometric distribution).

As z increases beyond 1, the only possible reason why $G(z)$ would become unbounded is that the denominator of the second term in (5.36) vanishes. Specifically

$$(1 - \rho) - \left\{ \frac{\lambda^2 E\{S^2\}}{2!} (z - 1) + \frac{\lambda^3 E\{S^3\}}{3!} (z - 1)^2 + \dots \right\} \to 0.$$

(5.38)

Since the series in brackets has all positive terms, it is a monotone increasing function of z for $z > 1$. For $\rho < 1$, there will be a $z_0 \geq 1$ where this expression vanishes. For ρ sufficiently close to 1, $z_0 - 1$ will be small and can be estimated by successive approximation or as a power series in $\rho - 1$. To second order in $\rho - 1$

$$z_0 - 1 \simeq \frac{2(1 - \rho)}{\lambda^2 E\{S^2\}} \left[1 - \frac{\lambda^3 E\{S^3\}(1 - \rho)}{3\lambda^4 E^2\{S^2\}} + \dots \right]$$

(5.39)

For $\rho \to 1$, the distribution of Q_s will approach a geometric distribution with a mean that depends only upon the first and second moments of S. Even if ρ is not very close to 1, it is possible to obtain estimates of the tail of the distribution.

5.8   The GI/G/1 system

The most obvious mathematical generalization of the M/G/1 system is the GI/G/1, a single-channel system with independent interarrival and service times each with some arbitrary distribution. If one chose to analyze the evolution of the queue length or $Q_s(t)$ in continuous time, then to predict the future stochastic behavior starting at an arbitrary time t one would need to specify not only the existing queue length but also the length of time since the last arrival and the time since the last customer entered service. Although this would generate a Markov process, it would be on a space with three state variables. One could simplify this by looking at the system only at certain discrete times, either an arrival time or a service completion time. At an arrival time, for example, one could describe the future behavior of the system if one knew only the queue length and the time since the last customer entered service. This would give a Markov process on a space with only two state variables.

The key step in the analysis of such a system is to pose the right question. If we consider the sequence of waiting times for successive customers w_j, we know from (1.10) that for FIFO queue discipline the

w_j satisfy the recursion relation

$$w_{j+1} = \max\ (0, w_j + s_j - \tau_{j+1}) \qquad (5.40)$$

in which s_j is the service time of the jth customer and τ_{j+1} is the time interval between the arrivals of the jth and $j+1$th customers. If the service time s_j of the jth customer and the time until the next arrival τ_{j+1} are statistically independent of w_j, then (5.40) would describe the distribution of w_{j+1} for any value of w_j. The w_j would define a Markov process on a one-dimensional space.

It is possible to derive explicit formulas for the equilibrium distribution of w, but the formulas are very clumsy to evaluate, except for special service and/or interarrival time distribution. The solution is obtained by solving an integral equation. Rather than consider the complete distribution of the w_j, we will consider only some simple properties of the first moments, but starting from a system even more general than the $GI/G/1$ system.

If $\tau_{j+1} - s_j - w_j > 0$, this quantity actually represents the length of time that the server is idle from the time the jth customer completes service until the next customer arrives. We denote this by I_{j+1} so that (5.40) can be written as

$$w_{j+1} - I_{j+1} = w_j + s_j - \tau_{j+1}, \qquad (5.41)$$

with $w_{j+1} \geq 0$, $I_{j+1} \geq 0$ and $w_{j+1} I_{j+1} = 0$, i.e., either $w_{j+1} = 0$ or $I_{j+1} = 0$.

Suppose that we have a sequence of $n+1$ customers chosen so that the system begins and ends in an empty state, i.e., $w_0 = 0$ and $w_n = 0$. If we sum equations (5.41) over j from $j = 0$ to $n-1$,

$$\sum_{j=0}^{n-1} w_{j+1} - \sum_{j=0}^{n-1} I_{j+1} = \sum_{j=0}^{n-1} w_j + \sum_{j=0}^{n-1} (s_j - \tau_{j+1}).$$

Since $w_0 = w_n = 0$ the sum of the w_j cancel and we have

$$\sum_{j=0}^{n-1} \tau_{j+1} = \sum_{j=0}^{n-1} s_j + \sum_{j=0}^{n-1} I_{j+1}. \qquad (5.42)$$

This is an obvious relation which merely says that the total time of observation is the sum of the service times plus the idle times.

We could also divide this by n and define appropriate arithmetic means so that this relation is written in the form

$$\langle \tau \rangle = \langle s \rangle + \langle I \rangle$$

or

$$\frac{\langle s \rangle}{\langle \tau \rangle} = \rho = 1 - \frac{\langle I \rangle}{\langle \tau \rangle}.$$

If for $n \to \infty$ the s_j, τ_{j+1}, I_{j+1} define some joint distribution we could also interpret arithmetic averages as expectations, i.e.,

$$\frac{E\{s\}}{E\{\tau\}} = \rho = 1 - \frac{E\{I\}}{E\{\tau\}}.$$

Suppose now that we square both sides of (5.41) and sum over j;

$$\sum_{j=0}^{n-1} (w_{j+1}^2 - 2w_{j+1}I_{j+1} + I_{j+1}^2) = \sum_{j=0}^{n-1} [w_j^2 + 2w_j(s_j - \tau_{j+1})$$
$$+ (s_j - \tau_{j+1})^2].$$

Since $w_0 = w_n = 0$, the sum of the w_j^2 cancel and $w_{j+1}I_{j+1} = 0$, so

$$\frac{1}{n} \sum_{j=0}^{n-1} I_{j+1}^2 = \frac{1}{n} \sum_{j=0}^{n-1} 2w_j(s_j - \tau_{j+1}) + \frac{1}{n} \sum_{j=1}^{n-1} (s_j - \tau_{j+1})^2.$$

If we interpret $E\{\cdot\}$ to mean arithmetic average, we can also write this as

$$E^2\{I_{j+1}\} + \mathrm{Var}\{I_{j+1}\} = 2E\{w_j\}E\{s_j - \tau_{j+1}\}$$
$$+ 2\,\mathrm{Cov}\{w_j, s_j - \tau_{j+1}\}$$
$$+ E^2\{s_j - \tau_{j+1}\} + \mathrm{Var}\{s_j - \tau_{j+1}\}.$$

From (5.42), $E^2\{I_{j+1}\} = E^2\{\tau_{j+1} - s_j\}$. If we solve this for $E\{w_j\}$,

$$E\{w_j\} = \frac{\mathrm{Var}\{s_j - \tau_{j+1}\} + 2\,\mathrm{Cov}\{w_j, s_j - \tau_{j+1}\} - \mathrm{Var}\{I_{j+1}\}}{2[E\{\tau_{j+1}\} - E\{s_j\}]}.$$

(5.43)

This is an algebraic identity valid for any sequence of w_j, s_j, τ_j with $w_0 = w_n = 0$.

The w_j appears on both sides of (5.43) but for the GI/G/1 system w_j, s_j and τ_{j+1} are assumed to be statistically independent, therefore uncorrelated. Actually, for any finite n the condition $w_0 = w_n = 0$ will induce some correlation between w_j and $s_j - \tau_{j+1}$ but, for sufficiently large n and $\rho < 1$, many of the w_j will vanish. This covariance term can be neglected. Also,

$$\mathrm{Var}\{s_j - \tau_{j+1}\} = \mathrm{Var}\{s\} + \mathrm{Var}\{\tau\},$$

so

$$E\{w_j\} = \frac{\rho[\text{Var}\{s\} + \text{Var}\{\tau\} - \text{Var}\{I\}]}{2E\{s\}(1-\rho)}.$$

(5.44)

The time average queue length can also be expressed as

$$E\{Q\} = \lambda E\{w_j\} = \frac{\rho^2 C^2(S) + C^2(\tau) - \lambda^2 \text{Var}\{I\}}{2(1-\rho)}.$$ (5.45)

In general, the distribution of the I_{j+1} may be quite complex and difficult to evaluate. If ρ is close to 1, however, most of the I_{j+1} will be zero and even if I_{j+1} is positive, it must be less than the corresponding τ_{j+1}. We would expect $\text{Var}\{I\}$ to be small compared with $\text{Var}\{s\}$ and/or $\text{Var}\{\tau\}$ for $1 - \rho \ll 1$. In any case $\text{Var}\{I\} \geq 0$, so

$$E\{w_j\} \leq \frac{\rho[\text{Var}\{s\} + \text{Var}\{\tau\}]}{2E\{s\}(1-\rho)},$$ (5.46)

and

$$E\{Q\} \leq \frac{\rho^2 C^2\{S\} + C^2(\tau)}{2(1-\rho)}.$$ (5.47)

In the special case of Poisson arrivals, we already know $E\{Q\}$ from (5.37). In this case, however, we can also evaluate $\text{Var}\{I\}$ because the time to the next arrival given that $I > 0$ is exponentially distributed independent of any previous events. The probability that an arriving customer finds the server idle (or a departing customer leaves the system idle) is $1 - \rho = P\{I > 0\}$. Therefore

$$E\{I\} = (1-\rho)/\lambda, \quad E\{I^2\} = (1-\rho)2/\lambda^2$$
$$\text{Var}\{I\} = (1-\rho^2)/\lambda^2.$$

Indeed, in this case, $\text{Var}\{I\} \to 0$ for $\rho \to 1$ but for $\rho \to 0$ the idle time is nearly equal to the interarrival time, $\text{Var}\{I\} \to \text{Var}\{\tau\} = 1/\lambda^2$. In (5.45)

$$C^2(\tau) - \lambda^2 \text{Var}\{I\} = 1 - (1-\rho^2) = \rho^2,$$

and

$$E\{Q\} = \frac{\rho^2[C^2(s) + C^2(\tau)]}{2(1-\rho)}$$

in agreement with (5.37).

Generally, we expect (5.46) or (5.47) to be a very close bound for $(1-\rho) \ll 1$ but a very poor one for $\rho \ll 1$.

Problems

5.1 Evaluate T_0, L_0, and $E\{Q_s\}$ for the random walk queue with $\lambda = 0.4$ and $\mu = 0.5$ as in problems 1.1 and 4.1. From the observations in problem 1.1 evaluate the fraction of time that $Q_s(t) = k$, $k = 0, 1, 2, \ldots$ over the range $t = 51$ to 150 and evaluate

$$\frac{1}{100} \sum_{t=51}^{150} Q_s(t).$$

5.2 For a single-channel server with independent service times S_j, the cumulative number of customers to arrive by time t is $A(t)$; the number to leave by time t is $D_q(t)$. For (arbitrarily) large t

$$\frac{\text{Var}\{A(t)\}}{E\{A(t)\}} \to I_A, \qquad \frac{E\{A(t)\}}{t} \to \lambda.$$

Show that

$$\frac{\text{Var}\{D_q(t)\}}{E\{D_q(t)\}} \to \begin{cases} \dfrac{\text{Var}\{S\}}{E^2\{S\}} & \text{if } \lambda E\{S\} > 1 \\[2mm] I_A & \text{if } \lambda E\{S\} < 1 \end{cases}.$$

5.3 For the M/M/m queue with $m^{1/2} \gg 1$, let

$$z = \frac{j - m}{m^{1/2}}, \qquad \kappa = m^{1/2}(1 - \rho).$$

For z and κ comparable with 1, show that

$$p_j \simeq \begin{cases} Am^{-1/2} \exp\left[-(z + \kappa)^2/2\right] & \text{for } z < 0 \\ Am^{-1/2} \exp\left(-\kappa^2/2 - \kappa z\right) & \text{for } z > 0 \end{cases}$$

with

$$A^{-1} \simeq (2\pi)^{1/2}\Phi(\kappa) + \kappa^{-1}\exp\left(-\kappa^2/2\right).$$

For the corresponding M/M/m/c,

$$p_c \simeq A_c m^{-1/2} \exp\left(-\kappa^2/2 - \kappa c^*\right)$$

with

$$A_c^{-1} \simeq (2\pi)^{1/2}\Phi(\kappa) + \kappa^{-1}\exp\left(-\kappa^2/2\right)(1 - e^{-\kappa c^*})$$

and

$$c^* = (c - m)/m^{1/2} > 0.$$

5.4 Suppose as in the random walk queueing system of Section 5.3 customers can arrive only at integer times with

$$P\{\text{arrival at time } t\} = \lambda, \quad P\{\text{no arrival at time } t\} = 1 - \lambda,$$

independent of any other events. The service times S_j, however, are independent with some arbitrary distribution on the integers $1, 2, \ldots$ ($E\{S\}$ and $\text{Var}\{S\}$ are finite). Determine $E\{Q_s\}$ and $E\{Q\}$ in terms of the moments of S.

CHAPTER 6

Independent or weakly interacting customers

6.1 Introduction

Nearly all of the textbook literature on queues deals with the application of the theory of Markov processes to queueing systems and, furthermore, is almost entirely restricted to the evaluation of equilibrium distributions. One exception to the above pattern is a class of systems in which customers do not physically interact or interact only occasionally or slightly while in the system under consideration. Most of these should probably be described as 'stochastic service systems' rather than 'queueing systems' since there is (usually) no queue, but they have customarily been included as a part of the literature of queueing theory, perhaps because such systems can be considered as limiting cases of queueing systems or because some people use the same mathematical techniques to analyze them as for systems with queues. In fact, the best techniques for analyzing such systems are quite different from those used in describing queues.

The classic example of a system with no queues is the infinite channel server; the $M/M/\infty$ or $M/G/\infty$ systems, for example. In many applications, however, there need not be physical entities which can be identified as discrete 'channels', or even something which one would normally consider to be performing a 'service'. Objects (customers) may simply be entering and leaving a system but not interfering with each other. Even if there are discrete channels (or some capacity constraint) there may be only a finite number, but enough so that the behavior of a customer is not affected by the presence of others; for example, people walking through an uncongested corridor, cars traveling along an uncongested freeway, or cars parked in a lot which is never (or seldom) full.

To model such systems, suppose there is only a finite number N

(possibly very large) of potential customers and we number them in some arbitrary order. Let T_k be the arrival time of customer k and S_k be his service time. We do not wish to make any assumptions yet about possible statistical dependencies among the T_j or between S_k and T_k. As a general model of noninteracting service times, we postulate that the joint distribution of the T_j, S_j has the product structure

$$P\{S_1 \le s_1, T_1 \le t_1, \ldots, S_n \le s_n, T_n \le t_n\}$$
$$= P\{S_1 \le s_1 | T_1 \le t_1\} P\{S_2 \le s_2 | T_2 \le t_2\} \ldots P\{S_n$$
$$\le s_n | T_n \le t_n\} P\{T_1 \le t_1, T_2 \le t_2, \ldots, T_n \le t_n\}, \qquad (6.1)$$

i.e., given all the T_j, the S_k are statistically independent with the conditional distribution of S_k dependent (at most) only on the value of T_k.

This postulate automatically excludes the usual systems in which queueing occurs since for such systems the time at which a customer enters the system (in this case the server itself) depends upon the time at which some service was completed and a server became free; i.e., there is a statistical dependence between some $T_j + S_j$ and a $T_k, k \ne j$.

We are primarily interested in how the probability distribution for the number $Q_s(t)$ of customers in the system at time t depends upon the arrival and service time distributions. If we let

$$Y_k(t) = \begin{cases} 1 & \text{if customer } k \text{ is in the service at time } t \\ 0 & \text{if customer } k \text{ is not in the service at time } t \end{cases} \qquad (6.2)$$

then

$$Q_s(t) = \sum_{k=1}^{N} Y_k(t). \qquad (6.3)$$

If customer k arrives at time T_k, he will leave at time $T_k + S_k$; therefore,

$$P\{Y_k(t) = 1\} = P\{T_k < t < T_k + S_k\}. \qquad (6.4)$$

Since $Y_k(t)$ is a function only of T_k and S_k, it follows from (6.1) that the conditional random variables $Y_k(t) | \{T_j\}$ given all the T_j are statistically independent. Furthermore, the conditional distribution of $Y_k(t) | \{T_j\}$ depends only upon T_k. Thus $Q_s(t) | \{T_j\}$ is the sum of independent 0–1 random variables, each of which depends upon only one of the T_k.

Whether or not it is difficult to evaluate the unconditional distribution of $Q_s(t)$ depends mostly upon how complex may be the probability structure of the arrival times T_k.

Depending upon the nature of the arrival process and service time distribution, $Q_s(t)$ might be a (multidimensional) Markov process. If, however, the service times have an arbitrary distribution function $F_S(t)$, one must specify (at least) the length of time each of the $Q_s(t)$ customers has been in service in order to describe the future behavior of $Q_s(t)$ as a Markov process. Each customer's residual time in service will depend on how long he has already been there, unless the service time distribution is exponential. Clearly this is not a simple approach except for the $M/M/\infty$ system or some simple generalization thereof.

6.2 Independent arrivals: the $M/G/\infty$ system

In Sections 4.1 and 4.4 we proposed that if customers have no opportunity to interact prior to reaching the server in question (in particular, they had not passed through a queue at some other server, and did not travel in bunches), then the T_k would likely be statistically independent, though not necessarily identically distributed. If this is true, in addition to (6.1), then any random variables associated with the history of the kth customer are statistically independent of anything associated with any other customer. In particular, the $Y_k(\tau_k)$ are statistically independent for any choice of times τ_k, and $Q_s(t)$ is the sum of independent 0–1 random variables.

It follows from (6.3) that

$$E\{Q_s(t)\} = \sum_k E\{Y_k(t)\} = \sum_k P\{Y_k(t) = 1\} \tag{6.5}$$

and, for the $Y_k(t)$ statistically independent,

$$\text{Var}\{Q_s(t)\} = \sum_k \text{Var}\{Y_k(t)\} = \sum_k P\{Y_k(t) = 1\}\left[1 - P\{Y_k(t) = 1\}\right]. \tag{6.6}$$

Since each term of (6.6) is less than the corresponding term of (6.5),

$$\text{Var}\{Q_s(t)\} < E\{Q_s(t)\}. \tag{6.7}$$

In some potential applications of (6.5) and (6.6) only a few customers contribute significantly to these sums for any value of t even though N may be very large. Any customer for which $P\{Y_k(t) = 1\} = 0$, of course, contributes nothing to either (6.5) or (6.6), or (6.3). In such a case it may be appropriate to evaluate these sums term by term, particularly if most or all of the customers who are likely to be in service at time t have identically distributed $Y_k(t)$.

If, for example,

$$P\{Y_k(t) = 1\} = \begin{cases} \rho(t) & \text{for } N' \text{ values of } k \\ 0 & \text{for all other } N - N' \text{ values of } k, \end{cases}$$

(at least for a certain range of t), then

$$E\{Q_s(t)\} = N'\rho(t), \quad \text{Var}\{Q_s(t)\} = N'\rho(t)[1 - \rho(t)]$$

and, furthermore, $Q_s(t)$ has a binomial distribution with parameters N' and $\rho(t)$. If there are two or three classes of customers for which $P\{Y_k(t) = 1\} > 0$, one can group terms of (6.5) and (6.6). The contribution to $Q_s(t)$ from each group of identical customers will have a binomial distribution, but $Q_s(t)$, being the sum of independent binomially distributed random variables with different parameters, will not itself have a binomial distribution.

A model of this type was used by G. N. Steuart (1974)† to investigate how the number of aircraft, $Q_s(t)$, at gate positions of a major airport depends upon the strategy for scheduling flights. At some airports which are major transfer terminals (notably Chicago and Atlanta) an airline will try to bring in a 'bank' of N' flights within a short time period (10–15 minutes) and dispatch them also within a short time period 40–50 minutes later. The purpose is to provide easy transfer of passengers and baggage between all flights in the bank.

Aircraft do not arrive or leave exactly on schedule. Although the deviation of one flight from its schedule may be correlated with the deviation of other flights from their schedules (weather conditions and queues at the runways affect all flights), it may be appropriate sometimes to postulate that the arrival and departure times (T_k and $T_k + S_k$) of different flights are statistically independent, i.e., the $Y_k(t)$ for flights in the same bank are statistically independent and (nearly) identically distributed.

The major difficulty in trying to model any real physical process is to evaluate the relevant parameters or functions in the model from observations. The more complex the model, the more tedious is the analysis of the data. Even with this simple model involving a bank of nearly identically distributed random arrivals and departures, one must determine experimentally the distribution of deviations from

---

† Steuart, G. N. (1974) Gate position requirements at metropolitan airports. *Transportation Science*, **8**, 169–89.

the schedule and the function $\rho(t)$. If one wishes to make predictions as to the consequences of various actions, one must further conjecture as to how these distributions will be affected by the strategy. In the airport gate analysis the question was how close one could schedule the banks without causing excessive fluctuation in gate occupancy due to overlap of the occupancy distributions of adjacent banks.

In this application, as in most applications of 'infinite-channel' systems, the number of channels (gate positions) is finite. One is not concerned so much with the consequences of queueing if it were to occur as with determining how many channels one needs so that it will not occur too often. A reasonable estimate of this could be obtained by evaluating $E\{Q_s(t)\}$ plus one or two standard deviations as a function of t.

In contrast with the above example in which only a few terms of (6.5) or (6.6) contribute significantly to the sum, there are many other applications in which the number of potential customers is very large, but the probability that any particular kth customer is in service at time t, $P\{Y_k(t) = 1\}$, is small, for all k. If this is so, then $\text{Var}\{Q_s(t)\} \simeq E\{Q_s(t)\}$. Furthermore, since $Q_s(t)$ represents a count of independent rare events, it will be Poisson distributed even though the individual customers may have different distributions of arrival times or service times. This follows from the same type of argument as given in Section 4.4.

If $Q_s(t)$ is Poisson distributed, it suffices to evaluate the first moment, but (6.5) is not usually the most convenient formula for evaluating it. Since the sum in (6.5) is independent of the order of numbering customers and consists of a sum of many small terms, it is appropriate to sum (integrate) the contributions from customers in the order of their arrival. If $E\{A(\tau)\}$ is the expected number of arrivals between some arbitrary time origin and time τ then the expected number of customers to arrive between time τ and $\tau + d\tau$ can be represented as $dE\{A(\tau)\}$. We can, if necessary, extend the definition of $E\{A(\tau)\}$ so that $E\{A(\tau)\}$ is negative for $\tau < 0$.

We have not excluded the possibility that the service time S_k of a customer could depend upon his arrival time T_k. Those customers who arrive at time τ may, therefore, have a service time distribution

$$F_S(s|\tau) = \text{Prob}\{S < s | \text{customer arrives at time } \tau\}. \quad (6.8)$$

Of those who arrive at time τ an expected fraction $[1 - F_S(t - \tau|\tau)]$ will still be in service at time t. The total expected number in service is, therefore

$$E\{Q_s(t)\} = \int_{-\infty}^{t} [1 - F_S(t - \tau | \tau)] dE\{A(\tau)\}$$

$$= \int_{0}^{\infty} [1 - F_S(x | t - x)] dE\{A(t - x)\}. \qquad (6.9)$$

If $E\{A(\tau)\}$ is differentiable with an arrival rate

$$\lambda(\tau) = \frac{d}{d\tau} E\{A(\tau)\}$$

then

$$E\{Q_s(t)\} = \int_{0}^{\infty} \lambda(t - x)[1 - F_S(x | t - x)] dx. \qquad (6.10)$$

In the special case in which all service times are equal to some number S, i.e.,

$$F_S(x | \tau) = \begin{cases} 1 & x > S \\ 0 & x \leq S \end{cases} \quad \text{for all } \tau$$

(6.1) gives

$$E\{Q_s(t)\} = E\{A(t)\} - E\{A(t - S)\}$$

the expected number of arrivals during the time interval $t - S$ to t. This is the same relation discussion in Section 3.7 equation (3.11) but with the deterministic $A_c(t)$ in Section 3.7 replaced by the expectation $E\{A(t)\}$. If λ is constant and $F_S(x | \tau)$ is independent of τ, (6.10) reduces to

$$E\{Q_s(t)\} = \int_{0}^{\infty} \lambda[1 - F_S(x)] dx = \lambda E\{S\}, \qquad (6.11)$$

consistent with the general relation $L = \lambda W$ discussed in Sections 1.4 and 1.5.

In most books on queueing theory, the fact that $Q_s(t)$ has a Poisson distribution is usually identified as a property of the $M/G/\infty$ system. The service times are customarily assumed to be not only independent of each other, but also independent of the arrival times, i.e., $F_S(x | \tau) = F_S(x)$ (an unnecessary restriction). The Poisson distribution for $Q_s(t)$ is usually derived by a direct enumeration of the probabilities for all arrival and service times that would lead to $Q_s(t) = j, j = 0, 1, \ldots$, given that the arrival process is a Poisson process.

That the $M/G/\infty$ system has a Poisson distribution for $Q_s(t)$ even for a time-dependent arrival rate (and a time-dependent service time distribution) is one of the very few 'exact' results in the queueing theory literature dealing with time-dependent phenomena.

Here we have argued more directly that $Q_s(t)$ will have a Poisson distribution if it is a rare event for any specific kth customer to be in the service at time t, for all k, but that the behavior of the kth customer is statistically independent of any other, and N is very large. These are essentially the same conditions that would generate a Poisson arrival process. In Sections 4.3 and 4.4 the Poisson process was derived under the hypothesis that it was a rare event for any kth customer to arrive during any finite (non-random) period of observation (and the behavior of different customers were independent), which implies that it is also a rare event for a customer having a service time that is finite with probability 1 to arrive within a service time prior to time t and consequently be in service at time t.

That the arrival process is a Poisson process is certainly sufficient to guarantee that $Y_k(t) = 1$ is a rare event. The present argument, however, gives somewhat more insight into possible deviations from a Poisson distribution for $Q_s(t)$ than a formal derivation based upon the assumption that the arrival process is exactly Poisson. Real processes that are only approximately Poisson often fail to be exactly Poisson because the number of arrivals over sufficiently long time intervals fails to have a Poisson distribution (for the reasons discussed in Section 4.5). From the above arguments it is clear that the arrival process need only behave like a Poisson process on a time scale comparable with the service time. Aside from possible statistical dependencies among the $Y_k(t)$, an obvious cause of possible deviations from a Poisson distribution for $Q_s(t)$ would arise if an appreciable fraction of the customers contributing to $Q_s(t)$ had a significant probability of being in service at time t (as in the airport gate occupancy example).

6.3   Multiple events

The properties discussed in the previous section are just special cases of a much wider class of properties for systems in which the history of events associated with any kth customer is independent of that associated with any other customer. The only events discussed in the last section were the events that a customer was in service at time t, i.e., $Y_k(t) = 1$. We could, however, identify many other possible events Z

associated with a kth customer and define $Y_k(Z) = 1$ if event Z happens to the kth customer, $Y_k(Z) = 0$ otherwise. Correspondingly, we can let

$$N(Z) = \sum_{k=1}^{N} Y_k(Z)$$

be the number of customers who experience the event Z.

In each case, $N(Z)$ is a sum of independent 0–1 random variables. If N is finite and the customers are not identical, the distribution of $N(Z)$ may be rather complicated. If, however,

$$P\{Y_k(Z) = 1\} = p_k(Z) = p(Z)$$

is independent of k, $N(Z)$ will have a binomial distribution with parameters $p(Z)$ and N. If $N \gg 1$ and $p_k(Z) \ll 1$ for all k, $N(Z)$ will have (approximately) a Poisson distribution with mean

$$E\{N(Z)\} = \sum_{k=1}^{N} p_k(Z).$$

If there are several events $Z_1, Z_2, \ldots$ that can happen, we can define counts $N(Z_1), N(Z_2), \ldots$. For example, Z_1 could be the event that a customer is in service at a time t_1 and Z_2 that he is in service at time t_2 or, if there is more than one server in some system of servers, Z_1 could be the event that a customer is in the first server, Z_2 that he is in the second server, etc.

Suppose some set of events $Z_1, Z_2, \ldots$ are mutually exclusive, i.e., if customer k experiences event Z_j he cannot also experience event Z_l, $j \neq l$. For example, if a customer is in server j he cannot simultaneously also be in server l. If the customers have identical stochastic behaviors, then the joint distribution of $N(Z_1), N(Z_2), \ldots$ will be multinomial with parameters $p(Z_1), p(Z_2) \ldots$ and N. If $N \gg 1$ and $p_k(Z_j) \ll 1$ for all k and j, then $N(Z_1), N(Z_2), \ldots$ will be (approximately) Poisson distributed and independent.

It follows immediately from this that if customers entering some network of servers form a (time-dependent) Poisson process and they do not interact while in the network (all servers are, in effect, infinite-channel servers) then the numbers of customers $Q_{s1}(t), Q_{s2}(t), \ldots$ in the services $1, 2, \ldots$ at time t are Poisson distributed and statistically independent. The $Q_{s1}(t)$ and $Q_{s2}(t')$, for $t \neq t'$, however, will not necessarily be statistically independent since the same customer could be in server 1 at time t and server 2 at time t' (the events in question are not mutually exclusive).

As a second example, suppose that one has N identical machines. The kth machine operates for a random time T_{kj} before failing, is repaired in a time S_{kj}, and then operates for a time $T_{k, \, j+1}$, etc. If there are no constraints on the number of machines which can be repaired or operate at the same time, the T_{kj}, S_{kj} are all statistically independent, and any starting conditions (if any) are also independent, then any history of events for the kth machine is independent of those for other machines. Since a machine cannot simultaneously be operating and be in repair, the number of machines operating (or in repair) at time t will be binomially distributed with some appropriate parameter $p(t)$. For sufficiently large t, the $p(t)$ will approach an equilibrium value equal to the long time fraction of time that a machine is operating, i.e.,

$$p(t) \to E\{T\}/E\{S+T\}.$$

Thus the equilibrium probability that m of the N machines are operating and $N-m$ are in repair is

$$\binom{N}{m} \frac{[E\{T\}]^m [E\{S\}]^{N-m}}{[E\{T\}+E\{S\}]^N}.$$

More generally, if one has any 'closed network' in which N identical customers circulate independently so as to spend certain fractions of time $p_1, p_2, \ldots, p_n$ in subsystems $1, 2, \ldots, m$ with

$$\sum_{j=1}^{n} p_j = 1,$$

the equilibrium number of customers in subsystems $1, 2, \ldots, n$ will be multinomially distributed.

If two events are not mutually exclusive, they can always be represented as the unions of mutually exclusive events. If, for example, one is interested in the joint properties of $Q_s(t')$ and $Q_s(t)$ for $t > t'$, we could let Z_1 be the event that a customer is in service at both times t' and t, Z_2 that he is in service at time t' but not t, and Z_3 that he is in service at time t but not t'.

The events Z_1, Z_2, and Z_3 are mutually exclusive and

$$Q_s(t) = N(Z_1) + N(Z_3)$$
$$Q_s(t') = N(Z_1) + N(Z_2). \tag{6.12}$$

If arrivals form a Poisson process $N(Z_1)$, $N(Z_2)$, and $N(Z_3)$ are independent Poisson distributed random variables with parameters $E\{N(Z_j)\}$, even if the service time of a customer may depend upon his arrival time.

Since $E\{N(Z_1)\}$ is the expected number of customers who are in service at both times t' and t, it is the sum (integral) of all those who arrived between times τ and $\tau + d\tau$, $\tau < t'$, and are still there at time t,

$$E\{N(Z_1)\} = \int_{-\infty}^{t'} [1 - F_S(t-\tau|\tau)] dE\{A(t)\}$$
$$= \int_{t-t'}^{\infty} [1 - F_S(x|t-x)] \lambda(t-x) dx. \quad (6.13)$$

In particular, if $F_S(x|t-x) = F_S(x)$ is independent of the arrival time and $\lambda(t) = \lambda$ is independent of t, then

$$E\{N(Z_1)\} = \lambda \int_{t-t'}^{\infty} [1 - F_S(x)] dx.$$

Similarly, the expected number of customers $E\{N(Z_2)\}$ that are in service at time t' but leave between time t' and t is

$$E\{N(Z_2)\} = \int_{-\infty}^{t'} [F_S(t-\tau|\tau) - F_S(t'-\tau|\tau)] dE\{A(\tau)\}$$
$$= \int_{t-t'}^{\infty} [F_S(x|t-x) - F_S(x-t$$
$$+ t'|t-x)] \lambda(t-x) dx. \quad (6.14)$$

The expected number of customers $E\{N(Z_3)\}$ who arrive during the time t' to t and are still in service at time t is

$$E\{N(Z_3)\} = \int_{t'}^{t} [1 - F_S(t-\tau|\tau)] dE\{A(\tau)\}$$
$$= \int_{0}^{t-t'} [1 - F_S(x|t-x)] \lambda(t-x) dx. \quad (6.15)$$

Since $N(Z_1)$, $N(Z_2)$, and $N(Z_3)$ are statistically independent and Poisson distributed, it follows immediately that

$$\text{Cov}\{Q_s(t), Q_s(t')\} = \text{Cov}\{N(Z_1) + N(Z_2), N(Z_1) + N(Z_3)\}$$
$$= \text{Var}\{N(Z_1)\} = E\{N(Z_1)\}. \quad (6.16)$$

In particular, for a stationary process, the correlation coefficient

$$\frac{\text{Cov}\{Q_s(t), Q_s(t')\}}{\text{Var}\{Q_s(t)\}} = \frac{\int_{t-t'}^{\infty} [1 - F_S(x)] dx}{\int_{0}^{\infty} [1 - F_S(x)] dx}$$

is a monotonic function of $t - t'$ that decreases from 1 at $t - t' = 0$ to 0 as $t - t' \to \infty$.

One can describe much more about the evolution of $Q_s(t)$. For example, the probability that $Q_s(t') = m$ and that all of these customers are still in service at time t is equivalent to the probability that m customers are in service at both t' and t, i.e., $N(Z_1) = m$, and that no (other) customers were in the service at time t' but left during t' to t, i.e., $N(Z_2) = 0$;

$$\frac{[E\{N(Z_1)\}]^m}{m!} \exp[-E\{N(Z_1)\}] \exp[-E\{N(Z_2)\}]$$
$$= \frac{[E\{N(Z_1)\}]^m}{m!} \exp[-E\{Q_s(t')\}].$$

Similarly, the probability that $Q_s(t) = m$ and that all of these customers have been in the service since t' is equal to the probability that $N(Z_1) = m$ and $N(Z_3) = 0$;

$$\frac{[E\{N(Z_1)\}]^m}{m!} \exp[-E\{Q_s(t)\}]$$

The conditional probability that $Q_s(t') = m$ (or $Q_s(t) = m$) and all m customers have been in service between time t' and t is

$$\left[\frac{E\{N(Z_1)\}}{E\{Q_s(t)\}}\right]^m \quad \text{or} \quad \left[\frac{E\{N(Z_1)\}}{E\{Q_s(t')\}}\right]^m \tag{6.17}$$

respectively. The former also has the interpretation of being the conditional probability that if m customers are in the system at time t', the time of the first service completion by one of these customers occurs after time t. The latter can be interpreted as the conditional probability that if m customers are in the system at time t the last arrival by one of these customers was before t'.

The former probability is independent of the number of customers who arrive after time t' (some of which could also depart before time t) and the latter probability is independent of the number of customers who depart before time t (some of which could have arrived after time t'). The probability of no arrivals in t' to t is

$$\exp\left[-\int_{t'}^{t} \lambda(\tau) d\tau\right] \tag{6.18}$$

and the probability of no departures in t' to t is

$$\exp\left[-\int_{-\infty}^{t} [F_S(t-\tau|\tau) - F_S(t'-\tau|\tau)]\, dE\{A(\tau)\} \right]. \quad (6.19)$$

In particular, for stationary arrivals and service times, the probability that no customers arrive or depart during the time t' to t given that there are m customers in the system at either t' or t is

$$\left[\frac{\int_{t-t'}^{\infty} [1 - F_S(x)]\, dx}{\int_{0}^{\infty} [1 - F_S(x)]\, dx} \right]^{m} \exp[-\lambda(t-t')]. \quad (6.20)$$

The derivative of this with respect to t' or t,

$$\left\{ [1 - F_S(t-t')]m + \lambda \int_{t-t'}^{\infty} [1 - F_S(x)]\, dx \right\}$$
$$\times \frac{\left[\int_{t-t'}^{\infty} [1 - F_S(x)]\, dx \right]^{m-1}}{[E\{S\}]^{m}} \exp[-\lambda(t-t')] \quad (6.21)$$

represents the probability density of the time to the next arrival or departure after time t' (the last arrival or departure before t). The first term in brackets represents the contribution due to a departure at time t' (an arrival at t) and the second term from an arrival at time t' (a departure at t). For sufficiently small $t - t'$ the two terms in the bracket are approximately m and $\lambda E\{S\}$.

We saw in the last section that $Q_s(t)$ is Poisson distributed for any $M/G/\infty$ system and that the stationary distribution of $Q_s(t)$ is determined by $\lambda E\{S\}$, independent of any other properties of the service distribution $F_S(x)$. The above formulas show, however, that the dynamic properties of $Q_s(t)$ depend upon the shape of the service distribution.

Although for small values of m, the evolution of $Q_s(t)$ clearly depends upon $F_S(x)$, as m increases, this sensitivity diminishes. If there are m customers in service at time t' (or t) with $m \gg 1$ then, for $t - t'$ small compared with a typical service time (specifically $F_S(t - t') \ll 1$), it is a rare event that any particular kth customer should leave (arrive) during the time t' to t. Since whether one person leaves (arrives) is independent of whether another does, the number of customers to leave (arrive) in $t - t'$ should have approximately a

Poisson distribution, i.e., the process of departures (arrivals) in t' to t should behave (locally) like a Poisson process of rate $m/E\{S\}$.

In (6.20) this behavior is illustrated by the fact that

$$
\left[\frac{\int_{t-t'}^{\infty} [1 - F_S(x)]\,dx}{\int_0^{\infty} [1 - F_S(x)]\,dx} \right]^m = \left[1 - \frac{\int_0^{t-t'} [1 - F_S(x)]\,dx}{E\{S\}} \right]^m
$$

$$
= \exp\left\{ m \ln\left[1 - \frac{\int_0^{t-t'} [1 - F_S(x)]\,dx}{E\{S\}} \right] \right\}
$$

For $F_S(t - t') \ll 1$ and $m \gg 1$, this becomes approximately

$$
\exp\left(-m(t - t')/E\{S\} \right),
$$

i.e., the time to the next departure (last arrival) is approximately exponentially distributed with rate $m/E\{S\}$.

From this we conclude that on those rare occasions in which $Q_s(t)$ reaches a value m sufficiently large that it is not likely to stay at or above m very long (short compared with a service time), the dynamic behavior of the $Q_s(t)$ is similar to the $M/M/\infty$ system with $\mu = 1/E\{S\}$.

If there is a nonzero probability that $S = 0$ (or more generally if for $m \gg 1$, $F_S(E\{S\}/m)$ is not small compared with 1), the short service time customers pass through the system quickly and would not likely be among the m customers present at time t' (or t). The time to the next arrival after t', however, is affected by the presence of such customers since the λ includes customers of all service times (even though some customers may leave almost immediately after they arrive).

To describe the behavior of a system with a significant fraction of customers having very short service times, it would be useful to separate the customers into (1) those with service times larger than some small number ε and (2) those with service times less than ε. The two types would behave independently and $Q_s(t)$ would be the sum of $Q_s^{(1)}(t)$ and $Q_s^{(2)}(t)$. The $Q_s^{(1)}(t)$ would behave as described above but $Q_s^{(2)}(t)$ would be a process with $\lambda^{(2)} E\{S^{(2)}\} \ll 1$. The latter process would be one having an arrival rate $\lambda^{(2)}$ not necessarily small compared with $\lambda^{(1)}$ but a $Q^{(2)}(t)$ consisting of unit pulses of sufficiently short duration that they seldom overlap.

6.4 Dependent arrivals

There are many types of physical systems in which discrete objects enter a system and remain there for random times, and which might be modeled by a process satisfying (6.1). For example, students enroll in a school and remain enrolled for a certain length of time, but one student's time in the system may be (nearly) independent of others. A passenger may board a bus but when (or where) he alights may be independent of others. For some of these systems, however, the arrival process may be quite complex because customers might, for example, have previously passed through another server or arrived in bunches.

For such systems the detailed behavior for the distribution of $Q_s(t)$ may also be complex. If $E\{Q_s(t)\}$ is large compared with 1, the distribution of $Q_s(t)$ is likely to be approximately normal, but $\mathrm{Var}\{Q_s(t)\}/E\{Q_s(t)\}$ may be appreciably different from 1, i.e., $Q_s(t)$ is not approximately Poisson distributed. To describe the behavior of such systems for the purpose of design it often suffices to have a reasonable estimate of $E\{Q_s(t)\}$ and some crude estimate of $\mathrm{Var}\{Q_s(t)\}$.

An estimation of $E\{Q_s(t)\}$ presents no problem because equations (6.5), (6.9), and (6.11) are valid regardless of any possible statistical dependencies among the arrivals. Equations (6.9) and (6.11), in particular, show that $E\{Q_s(t)\}$ depends only upon the arrival rate and the service time distribution. It is the $\mathrm{Var}\{Q_s(t)\}$ or the distribution of $Q_s(t)$ that may be sensitive to statistical dependencies.

The $\mathrm{Var}\{Q_s(t)\}$ can be decomposed into a part due to the random service times and a part due to the random number of arrivals. Suppose we denote by $Q_s(t)|A$ the number of customers in service at time t given the arrival process $A(\tau), \tau \leq t$, i.e., the $\{T_j\}$. Then we can write

$$\mathrm{Var}\{Q_s(t)\} = E_A\{\mathrm{Var}_S\{Q_s|A\}\} + \mathrm{Var}_A\{E_S\{Q_s|A\}\} \qquad (6.22)$$

in which E_A, Var_A refer to expectations and variances relative to the arrival process and E_S, Var_S relative to the service times.

The first term of (6.22) can be evaluated easily since

$$\mathrm{Var}_S\{Q_s|A\} = \sum_k P\{Y_k(t) = 1|T_k\}[1 - P\{Y_k(t) = 1|T_k\}]$$

and

$$P\{Y_k(t) = 1|T_k\} = 1 - F_S(t - T_k|T_k) \quad \text{for } T_k \leq t.$$

If we let $dA(\tau)$ represent the (random) number of arrivals between time τ and $\tau + d\tau$ then the sum over k can be written as

$$\text{Var}_S\{Q_s|A\} = \int_{-\infty}^{t} [1 - F_S(t - \tau|\tau)] F_S(t - \tau|\tau) dA(\tau).$$

Therefore,

$$E_A\{\text{Var}_S\{Q_s|A\}\} = \int_{-\infty}^{t} [1 - F_S(t - \tau|\tau)] F_S(t - \tau|\tau) dE\{A(\tau)\}$$

$$= E\{Q_s(t)\} - \int_{0}^{\infty} [1 - F_S(x|t - x)]^2$$

$$dE\{A(t - x)\}. \qquad (6.23)$$

The second term of (6.22) can be written as

$$\text{Var}_A\{E\{Q_s|A\}\} = \text{Var}_A\left\{\int_{0}^{\infty} [1 - F_S(x|t - x] dA(t - x)\right\}. \qquad (6.24)$$

We could approximate this integral by a sum of contributions from intervals of width $\delta > 0$, i.e.,

$$\int_{0}^{\infty} [1 - F_S(x|t - x)] dA(t - x)$$

$$\simeq \sum_{j=0}^{\infty} [1 - F_S(j\delta|t - j\delta)] [A(t - j\delta) - A(t - (j + 1)\delta)]. \qquad (6.25)$$

Particularly if the service time distribution is independent of starting time, i.e., $F_S(j\delta|t - j\delta) = F_S(j\delta)$, the factor $[1 - F_S(j\delta)]$ will be a decreasing function of j and go to zero for $j + \infty$. Therefore, (6.25) represents a weighted sum of (random) arrivals with arrivals from earlier times (larger j) receiving smaller and smaller weight.

The variance of the sum (6.25) can always be expressed in terms of the covariances of the contributions from the jth and kth intervals, summed over all j and k, but this would tend to obscure certain qualitative properties of (6.24). The most common type of correlations in real arrival processes are of short range, arising from interactions between adjacent arrivals. Customers may arrive in clusters or, on the other hand, short headways may be excluded. If there are long range correlations they are likely to be on a time scale larger than the typical service time and contribute little to (6.24).

If the strongest correlations in arrivals occur within a time which is short compared with the typical service time, and we choose δ comparable with the range of the correlations, then it might be reasonable to assume that the correlations between arrivals in the jth and kth intervals are negligible. The main contribution to the variance of (6.25) comes from the variance of arrivals in the jth interval, i.e., the covariance between the jth interval and itself. This would be exactly true, for any δ, if customers arrived in simultaneous batches, the batch arrival times formed a Poisson process, and the batch sizes were statistically independent.

If, as in Section 4.8, we assume that

$$\text{Var}\{A(t-j\delta) - A(t-(j+1)\delta)\} = I_A E\{A(t-j\delta) - A(t-(j+1)\delta)\}$$

and that I_A is (nearly) independent of the time $t-j\delta$ or δ, then

$$\text{Var}_A\{E_S\{Q_s(A)\}\} \simeq I_A \int_0^\infty [1 - F_S(x|t-x)]^2 dE\{A(t-x)\},$$

and

$$\frac{\text{Var}\{Q_s(t)\}}{E\{Q_s(t)\}} \simeq 1 + (I_A - 1)$$

$$\times \frac{\int_0^\infty [1 - F_S(x|t-x)]^2 dE\{A(t-x)\}}{\int_0^\infty [1 - F_S(x|t-x)] dE\{A(t-x)\}}. \quad (6.26)$$

We notice first that if the arrivals form a (time-dependent) Poisson process, the postulates used to derive (6.26) are exactly true for any arbitrarily small δ and $I_A = 1$; the second term of (6.26) vanishes. This is consistent with the results of Section 6.2 since, for Poisson arrivals, $Q_s(t)$ is Poisson distributed for all t and, therefore, $\text{Var}\{Q_s(t)\} = E\{Q_s(t)\}$.

Since the integrals in (6.26) are positive, $\text{Var}\{Q_s(t)\}/E\{Q_s(t)\}$ is greater or less than 1 accordingly as I_A is greater or less than 1. The integral in the numerator is always less than that in the denominator; consequently, (6.26) has a value between I_A and 1 (for either $I_A \geq 1$ or $I_A \leq 1$). A batch arrival process is a typical example of a process with $I_A > 1$ whereas evenly spaced arrivals would give $I_A = 0$.

To understand how the service time distribution affects (6.26) for $I_A \neq 1$, consider the special case in which $F_S(x|t-x) = F_S(x)$, and the arrival rate $\lambda(t-x)$ is (nearly) constant, at least over a length of time which includes most service times. The coefficient of $I_A - 1$ in (6.26) then simplifies to

$$\int_0^\infty [1 - F_S(x)]^2 \, dx \bigg/ \int_0^\infty [1 - F_S(x)] \, dx. \qquad (6.27)$$

If we let S and S' denote two independent random variables, each with a distribution function $F_S(x)$, $[1 - F_S(x)]^2$ can be interpreted as the probability that S and S' are both larger than x, i.e., min $(S, S') > x$. Then, (6.27) is equivalent to $E\{\min(S', S)\}/E\{S\}$.

The value of $E\{S\} - E\{\min(S, S')\}$ is a measure of the 'width' of the distribution of S much as the standard deviation or the quartile range, and

$$C_S^* = 1 - E\{\min(S, S')\}/E\{S\}$$

has an interpretation similar to the coefficient of variation (but with the additional feature that $0 \le C_S^* \le 1$). In the case where (6.27) holds, (6.26) can be written as

$$\frac{\text{Var}\{Q_s(t)\}}{E\{Q_s(t)\}} \simeq 1 + (I_A - 1)(1 - C_S^*). \qquad (6.28)$$

Contrary to what one typically expects of stochastic service systems, that an increase in the fluctuations of either the arrivals or the service will give a lower 'level of performance' here we see that (6.28) is a *decreasing* function of the service time fluctuations (as measured by C_S^*) if $I_A > 1$.

If one looks at a typical realization of the arrival and departure curves $A(t)$ and $D_s(t)$ for such a system, analogous to Figs 1.5 or 3.6(b), one can readily see why this is true. An arrival curve with batch arrivals $(I_A > 1)$ may have some large vertical steps. If the service times are all equal $(C_S^* = 0)$, then $D_s(t)$ will mimic the steps in $A(t)$ at a time S later, and the difference $Q_s(t)$ will suffer jumps from both curves. If, however, the service times are independent with $C_S^* > 0$ (particularly if some service times are nearly zero) the curve $D_s(t)$ of ordered departures will be smeared out. Whereas a large I_A means large vertical fluctuations in the curve of $A(t)$ versus t, a $C_S^* > 0$ means that the $D_s(t)$ curve has random horizontal displacements. These two effects can partially neutralize each other.

6.5 Loss systems with Poisson arrivals, exponential service time

Suppose that a server has only $m < \infty$ channels but, if a customer arrives and finds all servers busy, he either disappears (is lost) or is routed to some other server. He does not join a queue and wait for a server to become free. Equivalently, there may be m 'primary servers' but infinitely many secondary servers. A customer will use any one of the primary servers if one is free; otherwise, one of the secondary servers. There could even be a preferential ordering of all servers so that each new customer will go to the lowest numbered server that is free.

For such a system one is usually interested in the distribution of the number of customers $Q_{sm}(t)$ in the m primary server at time t or, equivalently, the distribution of $m - Q_{sm}(t)$, the number of idle servers, and in the expected rate at which customers are turned away. In the case in which all servers are ordered, m would be considered as a variable and $Q_{sm}(t)$ would represent the number of busy servers among the first m servers.

The study of such systems has a long history starting with the pioneering work of Erlang on telephone traffic. It is still discussed primarily in the context of telephone traffic. The customers are people who wish to use a group of m telephone trunk lines and the service time is the duration of a call. Customers who find all lines busy may be turned away (they may try again) or the call may be routed to some 'overflow' lines.

It is usually assumed that customer arrivals define a Poisson process. This is almost certain to be valid for telephone traffic since, typically, a very large number of potential customers have access to the same trunk lines, but the probability of any particular customer making a call during some time interval is independent of whether anyone else does. It is also found empirically that the duration of calls (service time) is approximately exponentially distributed.

The same type of model applies to many other systems for which the arrivals might not be a Poisson process, or the service times exponentially distributed. For example, a bus or elevator can hold only a finite number of passengers. If it becomes full, it will not stop for boarding passengers, but it will stop to let passengers off. The passengers who are by-passed will likely form a queue and wait, but not for the same vehicle. In a grocery store parking lot, newly arriving customers prefer to park as close to the store as possible; there is at least an approximate preferential ordering of parking spaces (servers). $P\{Q_{sm}(t) = m\}$ would represent the probability that a customer who arrives at time t would find all of the m best parking spaces full and be forced to use a less desirable space (or wait for a good space to become free).

We can consider any of the above systems as if there were infinitely many servers with identically distributed service times, but each new customer goes to the lowest numbered available server. If the arrival process is Poisson and the service times are independent, then the history of any customer is independent of others except for the identity of the server that he uses. We already know from Section 6.2 that, for the $M/G/\infty$ system, the total number of customers in the system $Q_s(t) = Q_{s\,\infty}(t)$ has a Poisson distribution at all times even if the arrival rate is time dependent. In particular, if the arrival rate is constant, $Q_s(t)$ is Poisson distributed with mean $\lambda E\{S\}$.

Unfortunately, most of the simple properties of $Q_s(t)$ do not have simple generalizations to $Q_{sm}(t)$ for $m < \infty$. In particular, the nonequilibrium properties of $Q_{sm}(t)$ are very complex (unless m is so large that $Q_{sm}(t)$ seldom exceeds m; m is, in effect, infinite).

If the service time is exponentially distributed with rate μ and we let

$$p_{j,\,m}(t) = P\{Q_{sm}(t) = j\}$$

be the probability that j of the first m servers are busy, then $p_{j,\,m}(t)$ satisfies a set of differential equations similar to (5.19), namely

$$\frac{dp_{j,\,m}(t)}{dt} = -(\lambda + j\mu)p_{j,\,m}(t) + \lambda p_{j-1,\,m}(t) + (j+1)\mu p_{j+1,\,m}(t),$$

(6.29)

but with

$$p_{m+1,\,m}(t) \equiv 0, \quad 0 \le j \le m.$$

Although these equations would be valid even if λ were time dependent, the time-dependent solutions of (6.29) would be quite complex. Even the transient solution of (6.29), i.e., the solution with λ and μ independent of t but with arbitrary initial conditions $p_{j,\,m}(0)$, is cumbersome. It is intuitively clear, however, that most customers who are in service at time t will have been served by time t plus one or two mean service times $(1/\mu)$, and possibly have been replaced by new customers who arrived during this time period. The natural unit of time for any transient behavior is, therefore, comparable with $1/\mu$ (except in the unlikely situation $\lambda \ll \mu$ in which even the first server is usually idle and will remain so for a time of order λ^{-1}).

The equilibrium solution of (6.29), however, is quite simple. As in (5.20),

$$p_{j,\,m} = p_{j,\,m}(\infty) = \frac{(\lambda/\mu)^j}{j!}\,p_{0,\,m} \quad j = 0, 1, \ldots, m, \quad (6.30)$$

i.e., the relative probabilities of various states $p_{j,\,m}/p_{k,\,m}$ are the same as for the Poisson distribution $(j, k \le m)$, but the normalization implies that

$$p_{j,\,m} = \frac{(1/j!)(\lambda/\mu)^j}{\sum\limits_{k=0}^{m} (1/k!)(\lambda/\mu)^k}, \quad 0 \le j \le m. \tag{6.31}$$

For fixed m, the distribution of Q_{sm} is a truncated Poisson distribution.

If the system is in a state $j < m$, a newly arriving customer will go to some kth server with $k \le m$, but if the system is in state m, a newly arriving customer must go to a server with $k > m$ (is lost if there are only m servers). Since $p_{m,\,m}$ represents the fraction of time that the first m servers are all busy, it also represents the probability that a customer who arrives at some random time will be lost (to the first m servers). The rate at which customers are lost is $\lambda p_{m,\,m}$. The $p_{m,\,m}$ given by (6.31) is usually called Erlang's first loss function.

If all servers are ordered and we let $K(t)$ denote the lowest number available server at time t, then the equilibrium distribution of K is

$$P\{K > m\} = p_{m,\,m} = \frac{(1/m!)(\lambda/\mu)^m}{\sum\limits_{k=0}^{m} (1/k!)(\lambda/\mu)^k} \tag{6.32}$$

$$= [1 + m(\mu/\lambda) + m(m-1)(\mu/\lambda)^2 + \dots + m!(\mu/\lambda)^m]^{-1} \tag{6.32a}$$

which satisfies the recurrence relation

$$\frac{1}{p_{m,\,m}} = 1 + m\left(\frac{\mu}{\lambda}\right)\frac{1}{p_{m-1,\,m-1}}.$$

From (6.32a) one can immediately verify that $p_{m,\,m}$ is a decreasing function of m that goes to zero for $m \to \infty$ (the sum diverges). This is particularly interesting in the commonly occurring situation for which $\lambda/\mu \gg 1$. In (6.32a) it is clear that $p_{m,\,m}$ decreases nearly linearly with m with slope μ/λ, at least for $m\mu/\lambda \ll 1$. It will differ appreciably from 1 when $m\mu/\lambda$ is comparable with 1, but for $\mu/\lambda \ll 1$ and $m\mu/\lambda$ comparable with 1,

$$1 + m(\mu/\lambda) + m(m-1)(\mu/\lambda)^2 + \dots \simeq 1 + (m\mu/\lambda) + (m\mu/\lambda)^2 + \dots$$
$$= (1 - m\mu/\lambda)^{-1}.$$

Therefore

$$P\{K > m\} \simeq \begin{cases} 1 - m\mu/\lambda & \text{for} \quad m\mu/\lambda < 1 \\ 0 & \text{for} \quad m\mu/\lambda > 1. \end{cases}$$

To this first approximation, K is nearly uniformly distributed over the interval $(0, \lambda/\mu)$; in a parking lot, the best available parking space is (almost) equally likely to be anywhere within a region containing the average number (λ/μ) of cars.

As a second approximation for $\lambda/\mu \gg 1$, we could use a normal approximation to the Poisson distribution, particularly for $m - \lambda/\mu$ of order $(\lambda/\mu)^{1/2}$, namely

$$p_{m,m} \simeq \frac{\phi\left(\dfrac{m - \lambda/\mu}{(\lambda/\mu)^{1/2}}\right)}{(\lambda/\mu)^{1/2}\,\Phi\left(\dfrac{m - \lambda/\mu}{(\lambda/\mu)^{1/2}}\right)}, \tag{6.33}$$

in which $\phi(x)$ and $\Phi(x)$ are the normal probability density and distribution functions, respectively.

Fig. 6.1 illustrates these approximations for $\lambda/\mu = 10$ and 100, i.e., $(\lambda/\mu)^{1/2}$ equal to 3.17 and 10. The solid line curve $P\{K > m\}$ gives a smooth curve approximation to the exact $p_{m,m}$ for $\lambda/\mu = 10$ (which is meaningful only for integer m). The solid line curve $P\{K = m\}$ is a smooth curve through the corresponding values of $P\{K = m\}$ for integer m. The broken line curves describe the first approximations for $P\{K > m\}$ and $P\{K = m\}$, the uniform distribution. The curve labeled normal is the approximation (6.33).

For $\lambda/\mu = 10$, $(\lambda/\mu)^{1/2} = 3.17$, none of these approximations are very accurate, but they do show the qualitative behavior. The two dotted curves show $P\{K > m\}$ and $P\{K = m\}$ for $\lambda/\mu = 100$, but the horizontal and vertical scales are $10\,m$ and $10\,P\{K = m\}$. These were drawn from (6.33) to illustrate the convergence to the uniform distribution.

6.6   Loss systems, general service times

We have seen that $Q_s(t)$ has a Poisson distribution for any time-dependent Poisson arrival process and independent service times with any distribution. Although the time dependence of $Q_{sm}(t)$ for $m < \infty$ could be quite complex even for an exponentially distributed service time, the equilibrium distribution of Q_{sm} for homogeneous Poisson

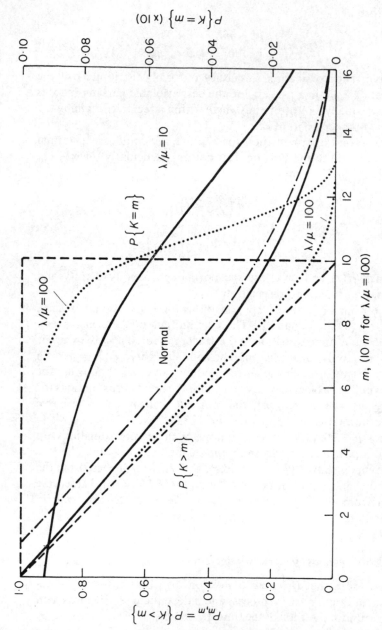

Figure 6.1 *Probabilities that the lowest ranked available server K is greater than m or equal to m*

arrivals is quite simple. The same is true for any service time distribution. Equation (6.31), with $1/\mu$ interpreted as $E\{S\}$, is valid for Poisson arrivals and independent service times with any distribution.

The proof that (6.31) is true for any service time distribution is not elementary. It is usually proved very formally from the equilibrium solution of a multidimensional Markov process defined on a state space representing the length of time each customer in service at time t has already been in service (since each customer's residual service time depends, generally, on how long he has already been there). To see why (6.31) is true at least for $m = 1$ or 2, consider a possible evolution of the occupancy of each server as illustrated in Fig. 6.2.

Each customer when he arrives goes to the lowest available server but remains in that server until his service is completed. Each line segment of Fig. 6.2 at height k means that the kth server is busy during that time. Its length is equal to some service time S_j sampled at random from the distribution $F_S(\cdot)$ but it can start only at a time when a customer arrives and finds all servers 1 to $k-1$ occupied. Obviously, a typical realization of this process contains some very complex patterns.

The formula (6.31) which we seek to prove does not describe the complete equilibrium joint probabilities for occupancies of all servers, i.e., the probability that server k is occupied for $k = 1, 2, \ldots$. It only describes the probability that j of the first m servers are busy (not which ones).

Consider first the case $m = 1$. We ask: What is the probability that the first server has $j = 0$ or 1 customers? From Fig. 6.2 one can see that the first server is busy for a time S_i and is then idle for a time T_i until the next customer arrives. The time T_i is exponentially distributed with rate λ and the S_j and T_j are independent of any past history of that server or what any other server is doing.

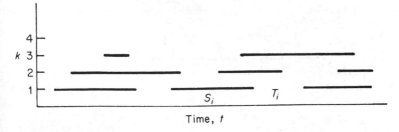

Figure 6.2 *Realization of service occupancies*

Over a long period of time containing $N \gg 1$ services the first server will be busy for a fraction

$$\frac{\sum\limits_{j=1}^{N} S_i}{\sum\limits_{i=1}^{N} (S_i + T_i)} = \frac{\frac{1}{N}\sum\limits_{j=1}^{N} S_i}{\frac{1}{N}\sum\limits_{i=1}^{N} (S_i + T_i)} \rightarrow \frac{E\{S\}}{E\{S\} + 1/\lambda} = \frac{\lambda E\{S\}}{1 + \lambda E\{S\}}.$$

This also represents the equilibrium probability $p_{1,1}$ that server 1 is busy in agreement with (6.31), i.e.,

$$p_{0,1} = \frac{1}{1 + \lambda E\{S\}}, \quad p_{1,1} = \frac{\lambda E\{S\}}{1 + \lambda E\{S\}}.$$

To evaluate $p_{j,2}$, consider first a hypothetical system consisting of two independent single-server loss systems, each serving a Poisson arrival process of rate λ with the same service time distributions. Let $Q'_{s2}(t) = 0$, 1, or 2, represent the number of servers that are busy at time t. Fig. 6.3(a) illustrates a possible realization of the occupancies of the two servers and Fig. 6.3(b) shows the corresponding realization of $Q'_{s2}(t)$.

From Fig. 6.3(a) it is clear that servers 1 or 2 are each busy a fraction $p_{1,1}$ of the time and are independent. Consequently, the probabilities that $Q'_{s2}(t)$ have values 0, 1, or 2 are

$$\frac{1}{(1 + \lambda E\{S\})^2}, \quad \frac{2\lambda E\{S\}}{(1 + \lambda E\{S\})^2}, \quad \text{and} \quad \frac{\lambda^2 E^2\{S\}}{(1 + \lambda E\{S\})^2}$$

respectively.

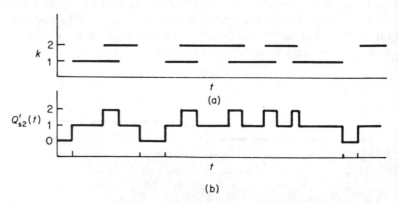

Figure 6.3   *Realizations of two independent single-channel loss systems*

Now compare the process $Q'_{s2}(t)$ with the process $Q_{s2}(t)$. Starting from any time when $Q'_{s2}(t)$ goes from 0 to 1 until it next returns to 0, the evolution of $Q'_{s}(t)$ is identical to a $Q_{s2}(t)$ starting from the same transition point. Both go from state 1 to 2 if a customer with arrival rate λ arrives before the first service is completed (otherwise to 0). It subsequently can go from state 2 to 1 when either service is completed, meanwhile rejecting any new arrivals, etc. The only difference between the processes is that $Q'_{s2}(t)$ and $Q_{s2}(t)$ remain in the state 0 for a time which is exponentially distributed the first with rate 2λ, the latter with rate λ.

From the process $Q'_{s2}(t)$ we can conclude that an idle period has a mean duration $1/(2\lambda)$ but that during a busy period $Q'_{s2}(t)$ must be in states 1 and 2 for an average total time of $E\{S\}$ and $\lambda E^2\{S\}/2$, respectively, in order for these times to be consistent with the above probabilities for $Q'_{s2}(t) = 0$, 1, or 2. The average times that $Q_s(t)$ spends in the states 0, 1, and 2 each cycle are therefore $1/\lambda$, $E\{S\}$, and $\lambda E^2\{S\}/2$. Therefore $p_{0,2}$, $p_{1,2}$, and $p_{2,2}$ are in the ratios 1, $\lambda E\{S\}$, and $[\lambda E\{S\}]^2/2$ as in (6.31).

Unfortunately these simple arguments for $m = 1$ and 2 are not easily generalized for $m > 2$. It is not obvious, but true, that the average length of time during a busy period that $Q_{sm}(t)$ is in state j is independent of m.

6.7   Bounds for the m-channel server

Since most practical queueing problems cannot be solved exactly, one must resort to various approximations. In some cases it is possible to obtain bounds on queue lengths, waiting times, etc., by showing that the system under consideration always has longer (or shorter) queues than some hypothetical system which can be analyzed exactly. Usually such bounds may be close under some conditions but not in others. In many cases it is not known how close the bounds are under any conditions.

Another approach is to seek approximations which are accurate only under certain limiting conditions. Most of the theory described in this book, for example, deals with approximations in which any relevant count of customers is so large that the customers can be treated as a continuous fluid. It is frequently possible also to analyze the opposite extreme, situations in which queues seldom exist and usually involve only a few customers at a time (perhaps 0 or 1). Such situations are seldom discussed in the queueing theory literature, perhaps because they are too elementary. If, however, one knows how a system behaves for very light traffic and for heavy traffic, one can sometimes make a gross interpolation and estimate at least the order of magnitude of queue lengths, delays, etc., in intermediate ranges of traffic where the detailed properties of the system are too complex to analyze in detail.

As an illustration of some bounds that could be useful for light traffic, we might try to exploit the fact that the behavior of the infinite-channel system can be analyzed under a wide variety of conditions, as described earlier in this chapter.

Suppose we have only $m < \infty$ identical servers and any customer who cannot enter service immediately must wait for a free server. Let $Q_{sm}(t)$ represent the number of customers in the system at time t (in service or waiting). If, on the other hand, any customer who was in the queue at time t were given service instead of just waiting and he could leave the system when he had obtained the requested service time from any server, this would certainly not increase the number of customers in the system at any later time. If, however, all customers were served while 'in the queue', the system would be equivalent to the infinite-channel server. Thus $Q_{sm}(t) \geq Q_{s\infty}(t)$. More generally, $Q_{sm}(t) \geq Q_{sm'}(t)$ for any $m' > m$ and for any realization for the arrival and service times (the same for the two systems).

We are concerned here mostly with situations in which a queue exists only rarely, but perhaps the existence of any queue is very costly, and one wishes actually to estimate how often it occurs, the queue lengths, waits, etc. In those cases discussed above in which the infinite-channel system behavior could be described, it would also, generally, be possible to determine the behavior of $Q_{s\infty}(t) - m$ when it is positive. In the case of the $M/G/\infty$ system, for example, even with the time-dependent arrival rates, $Q_{s\infty}(t)$ is Poisson distributed and the distribution of $Q_{s\infty}(t) - m > 0$ would be the tail of a Poisson distribution. The question is: how accurate an approximation is this for the queue length $Q_{sm} - m$ (particularly if $P\{Q_{s\infty}(t) - m > 0\} \ll 1$)?

There are two types of effects that cause $Q_{sm}(t)$ to differ from $Q_{s\infty}(t)$. First, if a customer in queue were to receive service while in queue, he might complete service and leave before he would otherwise have entered one of the m servers when it became free. This causes an immediate difference between $Q_{sm}(t)$ and $Q_{s\infty}(t)$ even if they had been nearly equal previously. This is potentially a significant source of differences between $Q_{sm}(t) - m$ and $Q_{s\infty}(t) - m$ if the waiting time of a customer in queue is comparable with his service time. For example, this is typically true for a single-channel server ($m = 1$) because the waiting time of a single customer in queue is usually comparable with the service time of the customer in service (and therefore with his own service time if all customers have the same service time distribution).

The second reason why $Q_{sm}(t) > Q_{s\infty}(t)$ is that, after a customer in queue enters the service, he stays there for a whole service time but, if he had received part of his service while in the queue, he will need only part of his service after one of the m servers becomes free. This, of course, causes future differences between $Q_{sm}(t)$ and $Q_{s\infty}(t)$. In particular, once $Q_{sm}(t)$ and $Q_{s\infty}(t)$ have dropped below m, it is more likely that $Q_{sm}(t)$ will exceed m again soon after, causing the queue to reform.

It is this last type of effect that makes exact solutions of most queueing problems difficult; any fluctuation initiates a 'chain reaction' of consequences which can persist for a long time. It is typical, however, of sufficiently light traffic that the consequences of a fluctuation which causes a queue to form persist only for a time which is short compared with the time until the next occurrence of a queue; each rare event can be analyzed separately and independently of others.

As an illustration of the differences between $Q_{sm}(t)$ and $Q_{s\infty}(t)$ we consider again the equilibrium M/M/m system (for which the exact distribution of queue length is known for all m), and the M/G/m (for which the exact distribution is known for $m = \infty$ and most properties can be determined for $m = 1$). Suppose that λ is constant and sufficiently low that $P\{Q_{sm}(t) - m > 0\} \ll 1$.

From Section 5.5, in particular equation (5.20), we already know that for the M/M/m system the equilibrium probabilities $p_{j,m} = P\{Q_{sm} = j\}$ satisfy the relations

$$\frac{p_{j+1,m}}{p_{j,m}} = \frac{\lambda}{(j+1)\mu} \quad \text{for} \quad j < m$$

$$\frac{p_{j+1,m}}{p_{j,m}} = \frac{\lambda}{m\mu} \quad \text{for} \quad j \geq m.$$

This distribution behaves like a Poisson distribution until $j = m$ but like a geometric distribution for $j \geq m$.

If $P\{Q_{sm} - m > 0\} \ll 1$ the normalization of the distribution for Q_{sm} is essentially the same as for $m = \infty$. Therefore,

$$p_{m,m} \simeq p_{m,\infty} = \frac{(\lambda/m\mu)^m}{m!} e^{-\lambda/m\mu},$$

but

$$P\{Q_{sm} - m = j\} = p_{m+j,m} = p_{m,m}\frac{1}{m^j}\left(\frac{\lambda}{\mu}\right)^j, \quad j \geq 0 \qquad (6.34)$$

whereas

$$P\{Q_{s\infty} - m = j\} = p_{m+j,\infty}$$
$$= p_{m,\infty}\frac{1}{(m+1)(m+2)\ldots(m+j)}\left(\frac{\lambda}{\mu}\right)^j.$$

As noted already in Section 5.5 and illustrated in Fig. 5.1, if $m \gg 1$ and $P\{Q_{sm} - m > 0\} \ll 1$, these two tail distributions will be nearly the same at least for the first few j values, and until $P_{m+j,m}$ is negligible compared with $p_{m,m}$ anyway. The bound $P\{Q_{s\infty} - m \geq j\} < P\{Q_{sm} - m \geq j\}$ is therefore close.

For the M/G/m system, $Q_{s\infty}$ is also Poisson distributed with $1/\mu$ interpreted as $E\{S\}$. Although we do not have an exact solution for the M/G/m system (the distribution for $Q_{sm} - m$ will not generally be exactly geometric), we expect the distribution of Q_{sm} to be nearly Poisson under essentially the same conditions as for the M/M/m system. For the M/G/m system, however, this may not be true for service distributions having a significant fraction of very short service times because customers with very short service times would pass quickly through an infinite-channel server but could be caught in a queue for the M/G/m system.

On the other hand, for small values of m, in particular for $m = 1$, (6.34) gives

$$p_{2,1} = p_{1,1}\lambda/\mu \simeq (\lambda/\mu)^2$$
$$p_{2,\infty} = p_{1,\infty}\frac{1}{2}\left(\frac{\lambda}{\mu}\right) \simeq \frac{1}{2}\left(\frac{\lambda}{\mu}\right)^2. \qquad (6.35)$$

Now the condition $P\{Q_{sm} - m > 0\} \ll 1$ implies also that $\lambda/\mu \ll 1$. The probability of having more than one customer in queue $p_{j,1}, j > 2$, is small compared with the probability for only one customer in the queue $p_{2,1}$. It is still true that $p_{2,\infty} < p_{2,1}$ but the bound is not very tight; $p_{2,\infty} \simeq (1/2)p_{2,1}$.

6.8 Successive approximations for small queues

To see why the bounds of the last section are not very close for small m, we consider now some more accurate approximations based upon the assumption that $\lambda(t)E\{S\} \ll 1$. Since, for stationary λ, $\rho = \lambda E\{S\}$

represents the expected number of busy servers, the servers must be idle most of the time. The logical approach is to look at the busy periods individually.

In order for two (or more) customers to be in the system simultaneously, it is necessary that some $(j+1)$th customer arrives within a service time S_j after a jth customer; if H_{j+1} is the headway (interarrival time) then $H_{j+1} < S_j$. Since for stationary arrivals $\lambda = 1/E\{H_j\}$, it follows, however, that $E\{H_{j+1}\} = 1/\lambda \gg E\{S_j\}$.

If customers arrive in batches there could be a significant probability that if one customer is in service there is more than one. The busy period can, in principle, be arbitrarily complex if customers arrive in batches with headways comparable with a typical service time. The 'batch' itself behaves as if one temporarily had a high arrival rate with potentially any structure of statistical dependencies. If, however, customers arriving in a batch have a headway short compared with a typical service time (virtually simultaneous arrivals) and $m < \infty$, it is fairly straightforward to describe the evolution of each batch. If (as is typically true for $\lambda E\{S\} \ll 1$) the system is empty when a batch arrives and the batch size exceeds m, a queue forms immediately. The queue will decrease by one after each service completion of the m busy servers until the queue vanishes. The server is likely to empty or be nearly empty before the next batch arrives.

Except in situations such as this in which there is a very strong dependence among the arrivals causing them to arrive virtually in batches, we expect that the condition $\lambda(t)E\{S\} \ll 1$ would imply that $P\{H_{j+1} < S_j\} \ll 1$. If most headways are comparable with $1/\lambda$ and most service times are comparable with $E\{S\}$ and we interpret ρ as a 'small parameter', then we can successively investigate effects of order $\rho^0 = 1, \rho^1, \rho^2, \ldots$.

In the 'zeroth approximation' the system is empty most of the time, i.e.,

$$p_{0,m}^{(0)} = 1, \quad p_{j,m}^{(0)} = 0, \quad j = 1, 2, \ldots.$$

In the next approximation we neglect events involving two or more customers in the system and let

$$p_{1,m}^{(1)}(t) = E\{Q_{s\,\infty}(t)\} = \int_0^\infty \lambda(t-x)[1 - F_S(x|t-x)]\,dx$$

as in (6.20). For stationary arrivals and service times

$$p_{1,m}^{(1)}(t) = \lambda E\{S\} = \rho.$$

Also
$$p_{0,m}^{(1)}(t) = 1 - p_{1,m}^{(1)}(t), \quad p_{j,m}^{(1)}(t) = 0, \quad j = 2, 3, \ldots.$$
To this approximation, the infinite-channel system, the loss system or any m-channel system $m \geq 1$ are equivalent under either stationary or time-dependent arrival patterns. There is no interference between customers because they seldom arrive close enough together to interact.

In the second approximation we neglect events involving three or more customers in the system simultaneously. To order ρ^2 only the single-channel system differs from the others. A customer who arrives while the single server is busy must wait until it is free.

The probability that a customer arrives between time $t - x$ and $t - x + dx$, $x > 0$ and is still in service at time t is
$$\lambda(t - x)[1 - F_S(x|t - x)]dx$$
(provided for $m = 1$ that no other customer was present at time $t - x$). Suppose we let $\lambda(t - x'|t - x)dx'$ represent the probability that, if a customer arrives at time $t - x$, another customer arrives between $t - x'$ and $t - x' + dx'$, $0 < x' < x$. If for a single-channel server both of these events occur, the second customer will also be in the system (in queue) at time t; there will be at least two customers in the system. Therefore
$$p_{2,1}^{(2)}(t) = \int_0^\infty \int_0^x \lambda(t - x)[1 - F_S(x|t - x)]\lambda(t - x'|t - x)dx'\,dx.$$
(6.36)
For $m > 1$, however, the second customer will still be in the system at time t only if his service time is greater than x', with probability $[1 - F_S(x'|t - x')]$. Thus
$$p_{2,m}^{(2)}(t) = \int_0^\infty \int_0^x \lambda(t - x)[1 - F_S(x|t - x)]\lambda(t - x'|t - x)$$
$$\times [1 - F_S(x'|t - x')]dx'\,dx \quad \text{for } m > 1.$$
(6.37)
In the special case $F_S(x|t - x) = F_S(x)$, and homogeneous Poisson arrivals $\lambda(t - x) = \lambda(t - x'|t - x) = \lambda$,
$$p_{2,1}^{(2)} = \lambda^2 \int_0^\infty [1 - F_S(x)]x\,dx = \lambda^2 E\{S^2\}/2,$$
but
$$p_{2,m}^{(2)} = \lambda^2 \int_0^\infty \int_0^x [1 - F_S(x)][1 - F_S(x')]dx\,dx'$$
$$= \lambda^2 E^2\{S\}/2 \quad \text{for } m > 1.$$

This already explains the differences in (6.35). For exponentially distributed S, $E\{S^2\} = 2E^2\{S\} = 2/\mu^2$. The phenomenon here is a familiar one; a second arrival is more apt to land in a long service time than a short one so his delay is proportional to $E\{S^2\}$ rather than $E^2\{S\}$.

For a general stationary arrival and service process $\lambda(t - x'|t - x)$ will be a function only of the time difference $x - x'$. To the approximation in which we neglect two or more arrivals within a service time

$$\lambda(t - x'|t - x) = f_H(x - x')$$

is essentially the probability density of the (marginal) headway distribution. If we let

$$F_H(z) = \int_0^z f_H(x)dx = P\{\text{headway} < z\}$$

then (6.36) simplifies to

$$p_{2,1}^{(2)} = \lambda \int_0^\infty [1 - F_S(x)]F_H(x)dx, \tag{6.38}$$

and (6.37) becomes

$$p_{2,m}^{(2)} = \lambda \int_0^\infty \int_0^x [1 - F_S(x)]f_H(x - x')[1 - F_S(x')]dx'dx$$

$$m > 1. \tag{6.39}$$

Equations (6.38) and (6.39) already show many of the qualitative features typical of the queue distribution at higher traffic intensities, particularly those relating to the interplay between the arrival and service distributions. In (6.38) and (6.39) the factors $[1 - F_S(x)]$ or $[1 - F_S(x')]$ will decrease with x or x' on a scale of x or x' comparable with $E\{S\}$. If $f_H(x)$ is nearly constant over this range (which is small compared with $E\{H\}$), i.e., $f_H(x) \simeq f_H(0)$ for x comparable with $E\{S\}$, then

$$p_{2,1}^{(2)} \simeq \lambda f_H(0)E\{S^2\}/2, \quad p_{2,m}^{(2)} \simeq \lambda f_H(0)E^2\{S\}/2 \quad \text{for} \quad m > 1. \tag{6.40}$$

For Poisson arrivals $f_H(0) = \lambda$. An $f_H(0) > \lambda$ would indicate a tendency for the arrivals to cluster (relative to a Poisson process), whereas $f_H(0) < \lambda$ would indicate a tendency for the arrivals to stay apart (as might happen, for example, if the arrivals were the output from another server). If it were virtually impossible for a headway to be less than a service time, $f_H(0) \ll \lambda$, there would, of course, hardly ever be more than one customer in service. Since a large $f_H(0)/\lambda$ would typically be associated with a large I_A, (6.40) indicates that both $p_{2,1}^{(2)}$ and $p_{2,m}^{(2)}$ ($m > 1$) increase with I_A as would be true also of the queue length at higher values of ρ.

In (6.40), $p_{2,1}^{(2)}$ increases with Var $\{S\}$ but $p_{2,m}^{(2)}$, $m > 1$ involves only the first moment of S. Actually $p_{2,m}^{(2)}$ as given by (6.39) will depend on the shape of the service distribution if $f_H(x)$ changes appreciably for $x \le E\{S\}$, and the dependence will be similar to that described in Section 6.3 for Var $\{Q_s\}$, particularly (6.26). If there were an excess of headways less than $E\{S\}$, i.e., a high value of $f_H(x - x')$ for x near x', its effect on (6.39) would be diminished by a low value of $[1 - F_s(x)]$ $[1 - F_S(x')]$ along $x = x'$. A broad distribution of S will be more effective than a narrow one to diminish the influence of an excess (or deficiency) of short headways on $p_{2,m}^{(2)}$.

We could also derive second order corrections to $p_{0,m}$ and $p_{1,m}$, but the details are not of much importance. The principal difference between $p_{0,m}$ for $m = 1$ and $m > 1$ comes from the fact that for $m = 1$ the single server must process all customers. In equilibrium $p_{0,1} = 1 - \lambda E\{S\}$ regardless of how large the queue may be, so any increase in $p_{2,1}$ comes at the expense of $p_{1,1}$. For $m = \infty$, however,

$$E\{Q_{sm}\} = p_{1,m} + 2p_{2,m} + \ldots = \lambda E\{S\}$$
$$\text{and } p_{0,m} + p_{1,m} + p_{2,m} + \ldots = 1.$$

To the approximation in which we neglect $p_{j,m}$ for $j > 2$ any increase in $p_{2,m}$ will also increase $p_{0,m}$, both at the expense of $p_{1,m}$.

To evaluate third, fourth, etc., order approximations for general time-dependent arrival processes becomes increasingly tedious. It is possible, however, to extend the above type of approximation so as to compare $p_{m+1,m}$ and $p_{m+1,\infty}$ for $m > 2$ and a Poisson arrival process. We are particularly interested to see why the bound $P\{Q_{sm} > k\} \ge P\{Q_{s\infty} > k\}$ becomes tighter as m increases as shown in (6.34) for the M/M/m system, and to show that it is true also for the M/G/m system.

6.9 The M/G/m system for light traffic

We will consider here the estimation of $p_{m+1,m}(t)$ only for stationary arrivals; the generalization to time-dependent arrivals is straightforward but gives somewhat clumsy formulas.

Suppose that λ is sufficiently small that $p_{m,m} \gg p_{m+1,m} \gg p_{m+2,m}$, etc. (which typically means that $\lambda E\{S\}/m \ll 1$). If this is true, then whenever the system reaches state $m+1$, it is almost certain to have come from state m rather than $m+2$. It is also likely to have reached state m from $m-1$, which means that $p_{m,\infty} \simeq p_{m,m}$. Indeed, $p_{j,\infty} \simeq p_{j,m}$ for all $j \leq m$.

For the infinite-channel system, $Q_{s\infty}(t)$ will have the value $m+1$ provided that: at some time $t - x - dx, x > 0$, the system was in state m, someone arrived during the time $t - x - dx$ to $t - x$ but no one arrived and none of the original m customers or the new one left during the time $t - x$ to t; or the system was in state $m+2$ at time $t - x - dx$, someone left during the time $t - x - dx$ to $t - x$ but no one arrived and none of the remaining $m+1$ customers left during the time $t - x$ to t. Under the assumed conditions, however, we can neglect the latter events and also neglect the possibility of an arrival in $t - x$ to t (which is much less likely to occur than a departure).

From (6.20) or (6.21) we conclude that

$$p_{m+1,\,\infty} \simeq p_{m,\infty} \int_0^\infty \left[\frac{\int_x^\infty [1 - F_S(y)]dy}{E\{S\}} \right]^m [1 - F_S(x)]\lambda dx$$

$$= p_{m,\infty} \lambda E\{S\}/(m+1). \tag{6.41}$$

(Actually, this final expression is exactly correct, since $Q_{s\infty}(t)$ is known to have a Poisson distribution; the neglect of transitions from state $m+2$ and of new arrivals after time $t - x$ have exactly cancelling effects.)

For the corresponding m-channel system, however, the customer who arrives at $t - x$ is certain still to be there at time t if none of the original m customers leaves during the time t to $t - x$. Therefore

$$p_{m+1,\,m} \simeq p_{m,\infty} \int_0^\infty \left[\frac{\int_x^\infty [1 - F_S(y)]dy}{E\{S\}} \right]^m \lambda dy$$

$$= p_{m,\infty} \int_0^\infty \left[1 - \frac{\int_0^x [1 - F_S(y)]dy}{E\{S\}} \right]^m \lambda dx. \tag{6.42}$$

This differs from (6.41) in that a factor $[1 - F_S(x)]$, the probability that the new customer arriving at time $t - x$ would have left by time t, is missing in (6.42).

In the special case of exponentially distributed service times $1 - F_S(y) = \exp(-\mu y)$, (6.41) and (6.42) give

$$\frac{p_{m+1,m}}{p_{m+1,\infty}} = \frac{m+1}{m}$$

in agreement with (6.24). For general service times but $m = 1$ these formulas give

$$\frac{p_{2,1}}{p_{2,\infty}} = \frac{E\{S^2\}}{E^2\{S\}}$$

in agreement with (6.36) to (6.40).

Particularly for $m \gg 1$; the fractional difference between (6.41) and (6.42) will be small because most of the contribution to the integral will come from small values of x for which $F_S(x) \ll 1$. It is convenient, therefore, to consider

$$\frac{p_{m+1,m} - p_{m+1,\infty}}{p_{m+1,\infty}} = \frac{(m+1)}{E\{S\}} \int_0^\infty F_S(x) \left[1 - \frac{\int_0^x [1 - F_S(y)]dy}{E\{S\}} \right]^m dx.$$

$$(6.43)$$

Except for $m = 1$, (6.43) cannot be expressed in terms of finite moments of S but, for $m \gg 1$, the integrand of (6.43) can be approximated by

$$\frac{p_{m+1,m} - p_{m+1,\infty}}{p_{m+1,\infty}} \simeq \frac{(m+1)}{E\{S\}} \int_0^\infty F_S(x) \exp\left(-\frac{mx}{E\{S\}} \right) dx.$$

$$(6.44)$$

If, in addition, the service times have a finite probability density at 0 so that

$$F_S(x) \simeq f_S(0)x \quad \text{for} \quad x \lesssim E\{S\}/m$$

(6.44) simplifies further to

$$\frac{p_{m+1,m} - p_{m+1,\infty}}{p_{m+1,\infty}} \simeq \frac{f_S(0)E\{S\}}{m} \quad \text{for} \quad m \gg 1. \qquad (6.45)$$

For an exponentially distributed service time $f_S(0) = 1/E\{S\} = 1/\mu$, (6.44) becomes $1/m$. If there is a deficiency of short service time $f_S(0)E\{S\} \ll 1$, the difference between $p_{m+1,m}$ and $p_{m+1,\infty}$ could be very small because it is unlikely that a customer who caused the transition to state $m + 1$ of an infinite-channel service would complete service before one of the other m servers became free.

On the other hand, if there is an excess of short service times, possibly some zero service times so that $F_S(0_+) \neq 0$, (6.44) becomes

$$\frac{p_{m+1,m} - p_{m+1,\infty}}{p_{m+1,\infty}} \simeq \frac{(m+1)}{m} F_S(0_+) \simeq F_S(0_+)$$

which means that $p_{m+1,m}/p_{m,\infty} \simeq 1 + F_S(0_+) > 1$ even for $m \gg 1$. In this extreme situation, customers with short (zero) service time would pass through an infinite-channel server quickly but would wait in queue like anyone else for $m < \infty$.

For $m \gg 1$ the assumption that $p_{m,m} \gg p_{m+1,m} \gg p_{m+2,m}$, etc. (or $\lambda E\{S\}/m \ll 1$) is quite severe because, if this were true, $p_{m,m}$ itself is probably negligibly small. For the $M/G/\infty$ system, $p_{m,\infty}$ is approximately normal for $\lambda E\{S\} \gg 1$ with mean $\lambda E\{S\}$ and variance $\lambda E\{S\}$. For the m-channel system, queueing would be unlikely unless $m - \lambda E\{S\}$ were less than about two standard deviations of $Q_{s\infty}$ i.e.,

$$\lambda E\{S\}/m \lesssim 1 - 2[\lambda E\{S\}]^{1/2}/m \simeq 1 - 2/m^{1/2}.$$

If queueing is not negligible for $m \gg 1$, the distribution of the queue Q_{sm} given that $Q_{sm} > m$ extends over several integer values (not just 1).

As explained in Section 6.3, however, the time-dependent behavior of $Q_{s\infty}(t)$ for the $M/G/\infty$ system is very similar to the $M/M/\infty$ system when $Q_{s\infty}(t)$ reaches some value m sufficiently large that it will likely exceed m only for some time x such that $F_S(x) \ll 1$. The only difference between the time-dependent behavior of $Q_{sm}(t)$ and $Q_{s\infty}(t)$ is that when $Q_{s\infty}(t) > m$, $Q_{s\infty}(t)$ servers are working but if $Q_{sm}(t) > m$ only m servers are busy. We expect, however, that $[Q_{sm}(t) - m]/m$ is small compared with 1 most of the time.

Problems

6.1 Exactly N applicants for a driver's license are told to take a written examination today at the Department of Motor Vehicles. A jth person will arrive at time T_j, pick up the examination, answer the questions, return the answers, and leave at time $T_j + S_j$.

If T_j and S_j are all independent random variables with distribution functions

$$F_T(\tau) = P\{T_j \le t\}$$
$$F_S(s) = P\{S_j \le s\} \quad 1 \le j \le N$$

derive a formula for the probability $P_k(t)$ that there are k applicants in the building at time t in terms of $F_T(\tau)$, $F_S(s)$ and N.

If $N = 320$, T_j is uniformly distributed over the time 9.00 a.m. to 5.00 p.m., and S_j is uniformly distributed over the time interval $1/4$ to $1/2$ hours, evaluate $p_k(t)$. Describe the approximate shape of this $p_k(t)$ as a function of k, and evaluate its expectation and variance.

6.2 Airplanes are scheduled to arrive at a terminal at times k (every hour on the hour) and leave at times $k+1$, $k =$ integer, $-\infty < k < \infty$. The kth aircraft, however, actually arrives at time

$$T_k = k + L_k$$

with L_k exponentially distributed and independent

$$P\{L_k > x\} = e^{-\beta x},$$

and it leaves at time $k+1$ or $T_k + 1/2$ whichever is later. There is no restriction on the number $Q_s(t)$ of aircraft that can be at the terminal at any time.

Evaluate $E\{Q_s(t)\}$, $\text{Var}\{Q_s(t)\}$ for $0 < t < 1$ and show that $Q_s(t)$ becomes Poisson distributed for $\beta \to 0$.

6.3 An infinite-channel server has independent identically distributed service times with distribution function $F_S(s) = P\{S \le s\}$ and the arrivals form a Poisson process. Determine the probability that no customer leaves the service during the time interval 0 to τ if

(a) the arrival process has a constant rate λ
(b) the arrival rate $\lambda(t)$ at time t is time dependent.

6.4 At time $t = 0$, there are n customers in an infinite-channel service. Each of the servers has exponentially distributed service time with rate μ. If no new customers arrive for $t > 0$, determine

(a) the probability that there are k customers in service at time t, $t > 0$,
(b) the expected number of customers in service at time t,
(c) the distribution function of the time until the first service completion after time 0,

(d) the distribution function of the time until the service is empty.

If there are n customers in service at time 0 but new customers arrive at a rate λ, what is

(e) the expected number of customers in service at time t?

6.5 Evaluate C_S^* in (6.28) if

(a) $P\{S > s\} = (1 - p)e^{-\mu s}, \quad 0 < p < 1,$
(b) $P\{S > s\} = \exp(-as^{\alpha}), \quad 0 < a, 0 < \alpha.$

Suppose that $p \to 1$ and $\mu \to 0$ or $\alpha \to 0$ and $a \to \infty$ with $E\{S\}$ fixed. Show that $\operatorname{Var} Q_s\{(t)\}/E\{Q_s(t)\} \to 1$, independent of I_A. Why does this system behave as if the arrivals formed a Poisson process?

6.6 Suppose, as in problem 4.8, that three airplanes are scheduled to arrive at $t = 0$ but the actual arrivals T_j^* are independent identically distributed random variables with a distribution function

$$F_T(t) = P\{T_j^* \le t\} = \begin{cases} 1 - e^{-t/\tau} & \text{for } t > 0 \\ 0 & \text{for } t < 0 \end{cases}$$

If there is a minimum time β between landings, determine $p_j(t)$ exactly. Reasonable values of τ and β would be 10 minutes and 2 minutes respectively.

Compare $p_j(t)$ with $p_{j,1}^{(0)}$, $p_{j,1}^{(1)}$, and $p_{j,1}^{(2)}$, of Section 6.8.

CHAPTER 7

Diffusion equations

7.1 Introduction

In order to analyze further the properties of equilibrium and time-dependent queues when the number of customers in the system is so large that the integer nature of the count can be ignored, we make use of some techniques that have been exploited by physicists and mathematicians for a couple of centuries.

The historical evolution of many physical laws, such as those describing gravitational and electrical fields, heat conduction, diffusion, fluid dynamics, etc., has typically begun with an experimental determination of some physical principle in an idealized geometry; for example, the field produced by a point mass or point electric charge, the temperature produced by a heat source, or the propagation of a simple wave. From these 'solutions', partial differential equations were derived; Laplace's equation for static potentials, Maxwell's equation for electromagnetic theory, the diffusion equation, wave equation, etc. Later the logic was reversed; the differential equation was considered to represent the 'fundamental' principle. Many textbooks now present painful derivations of the elementary solutions for partial differential equations, including the solution from which the differential equation was originally discovered.

There are several advantages to describing a physical law in the form of a differential equation. First of all, the differential equation displays a local cause and effect relation; an effect at some (vector) position x at time t is considered to arise from causes in the immediate neighbourhood (at points $x + dx$) at a slightly earlier time $t - dt$. Secondly, the differential equation is very convenient for the mathematical treatment of boundary conditions in the physical space or initial conditions (reflection of waves from a wall, absorption of particles on a surface, temperature distribution generated from an arbitrary initial distribution, etc.) because the boundary conditions or initial conditions do not appear explicitly in the differential equation;

they appear as separate conditions. Finally, if the differential equation is linear, one can generate multidimensional families of solutions by taking linear combinations of other solutions. A vast store of techniques for solving these equations has evolved from the interplay between logic developed from physical reasoning and from mathematical reasoning.

The behavior of many queueing systems is abstractly very similar to the behavior of certain physical systems, particularly diffusion or heat conduction in an external force field. In queueing theory we are usually interested in the evolution of some probability distributions; for example, the distribution of the queue length $Q(t)$ or waiting time w_j, or the joint distributions of the cumulative arrivals and departures $A(t)$, $D(t)$, or the arrival and departure times $A^{-1}(j)$, $D^{-1}(j)$. In any case the 'solution' is a (probability) function defined on some finite dimensional space. The space might actually be a lattice of integer valued variables, but in the fluid approximation we would treat the space as a continuum and might even describe the probability distribution in terms of a probability density. It is, of course, quite common in the probability literature to describe a probability density as a 'mass density' (with a unit total mass); the first moment of the probability density is the analogue of the center of gravity of the mass distribution and the variance (or covariance matrix) is the analogue of the moment of inertia about the center of gravity. The evolution of a probability density can correspondingly be identified as a movement of fluid in some space.

The motion of a physical fluid satisfies certain rules which presumably are universally valid, i.e., can be experimentally reproduced. In queueing theory we analyze the behavior of various hypothetical systems which may or may not correspond to some actual physical system. As a mathematical exercise we are free to imagine any behavior we like, but most systems of practical concern would have some local cause and effect behavior. If $A(t_0)$ and/or $D(t_0)$ were known at some time t_0, then at a slightly later time t, $A(t)$ and $D(t)$ will not have changed very much. Correspondingly, if one had some joint probability density for $A(t_0)$, $D(t_0)$ at time t_0, $f_{AD}(x_0, y_0; t_0)$, then the probability density at x, y at time t, $f_{AD}(x, y; t)$ would be related somehow to values of $f_{AD}(x_0, y_0; t_0)$ for x_0, y_0 in some neighborhood of x, y.

It is convenient to imagine that $f_{AD}(x, y; t)\,dx\,dy$ represents an amount of fluid in an element of area $dxdy$ in an (x, y) space as in Fig. 1.6(a) but to realize that it actually has the interpretation of being the fraction of all realizations (curves) $A(t)$, $D(t)$ which lie in $dxdy$ at time t. The local cause and effect behavior simply means that the customers do not arrive or leave in large batches so as to cause the curves $A(t)$, $D(t)$ to have sizeable jumps and also that the cause of the local motion is unrelated to past history. Of course all of the arguments here relating to what is 'small' or 'large' presuppose that the integer nature of $A(t)$, $D(t)$ can be ignored. The $dxdy$ does not really mean an 'infinitesimal' area but an element of area containing at least one lattice point, and a 'short time' usually means at least a few interarrival or service times, for otherwise the motion would depend upon the detailed partial service times of customers.

The main conjecture which we will exploit here is that during a time interval t_0 to t involving many arrivals and departures the changes in $A(t)$, $D(t)$ both behave approximately like a Brownian motion and are (nearly) independent of each other as discussed in Chapter 4, provided that $Q(t)$ (almost certainly) stays positive during the time interval.

With this assumption, we already know the 'solution' describing the evolution of the queue, for any length of time, provided that the queue remains positive. The troublesome point which we have avoided so far is how to handle the boundary condition that $Q(t)$ cannot be negative, i.e., the fluid cannot cross the line $x = y$ in Fig. 1.6(a). It is precisely for reasons analogous to this (to handle boundary and/or initial conditions) that physicists introduced partial differential equations to describe physical laws even though the differential equation may have been derived from certain elementary solutions.

7.2 The diffusion equation

We will derive here a two-dimensional version of the diffusion equation under somewhat more general conditions than necessary for the description of the process $A(t)$, $D(t)$ discussed in the last section (generalizations to higher dimensions will be obvious).

If

$f(x, y; t)$
   = joint probability density of $A(t)$, $D(t)$ (or any other pairs of random variables) at time t,

$f(x, y; t | x_0, y_0; t_0)$
   = conditional probability density of $A(t)$, $D(t)$ at location x, y and time t given that $A(t_0) = x_0$, $D(t_0) = y_0$,

then

$$f(x, y; t) = \iint f(x, y; t \,|\, x_0, y_0; t_0) f(x_0, y_0; t_0) \, dx_0 \, dy_0. \qquad (7.1)$$

The basic qualitative property of the process in question is that $A(t)$, $D(t)$ can change only by a small amount in a short time (even if the system is near a boundary). What this means relative to (7.1) is that for $t - t_0$ sufficiently small, the conditional density considered as a function of x, y is highly concentrated in some neighborhood of the point x_0, y_0.

This also implies that the conditional density considered as a function of x_0, y_0 is concentrated in some neighborhood of x, y. If $f(x, y; t \,|\, x_0, y_0; t_0)$, as a function of x, y, x_0, y_0 depended only upon $x - x_0$, $y - y_0$, then this function would also be a probability density relative to x_0, y_0 for any x, y, i.e.,

$$\iint f(x, y; t \,|\, x_0, y_0; t_0) \, dx_0 \, dy_0$$

$$= \iint f(-x_0, -y_0; t \,|\, -x, -y, t_0) \, dx_0 \, dy_0 = 1.$$

This is actually a logical postulate for the $A(t)$, $D(t)$ process discussed in the last section (provided x, y is not too close to a boundary). It simply means that the changes $A(t) - A(t_0)$, $D(t) - D(t_0)$ are independent of $A(t_0)$ and $D(t_0)$, but in particular of $Q(t_0) = A(t_0) - D(t_0)$ for $Q(t_0) > 0$.

In this case, a logical way to approximate the integral in (7.1) would be to suppose that $f(x_0, y_0; t_0)$ were a smooth function which for x_0, y_0 in some neighborhood of x, y could be approximated by a Taylor series expansion around x, y, i.e.,

$$f(x_0, y_0; t_0) \simeq f(x, y; t_0) + (x_0 - x) \frac{\partial f}{\partial x} (x, y; t_0)$$

$$+ (y_0 - y) \frac{\partial f}{\partial y} (x, y; t_0) + \frac{(x_0 - x)^2}{2} \frac{\partial^2 f}{\partial x^2} (x, y; t_0) + \ldots.$$

Now, relative to the integration variables x_0, y_0, $f(x_0, y_0; t_0)$ is approximately a polynomial. A term-by-term integration with respect to x_0, y_0 involves various moments of the conditional random variables $A(t_0) - A(t)$, $D(t_0) - D(t)$ multiplied by corresponding derivatives of $f(x, y; t_0)$. Thus $f(x, y; t)$ can be related to $f(x, y; t_0)$ and its derivatives $\partial f(x, y; t_0)/\partial x$, etc., at the earlier time t_0.

In the more general case in which the distribution of $A(t) - A(t_0)$, $D(t) - D(t_0)$ may depend upon $A(t_0)$, $D(t_0)$ (even if only near a boundary), $f(x, y; t|x_0, y_0; t_0)$ will always be a probability density relative to x, y for any x_0, y_0. To exploit this, it is advantageous to create some integrals relative to x, y. Let $g(x, y)$ be some arbitrary smooth function of x, y, and consider the integral

$$\iint g(x, y) f(x, y; t) \mathrm{d}x \, \mathrm{d}y = E\{g(A(t), D(t))\}$$

$$= \iiiint g(x, y) f(x, y; t|x_0, y_0; t_0) f(x_0, y_0; t_0) \mathrm{d}x_0 \mathrm{d}y_0 \mathrm{d}x \mathrm{d}y.$$

$$(7.2)$$

Now reverse the order of integration and consider the conditional density as a function of x, y for fixed x_0, y_0. Since the conditional distribution is concentrated near x_0, y_0, we need only consider values of $g(x, y)$ for x, y in some neighborhood of x_0, y_0. Suppose that $g(x, y)$ can be approximated by a Taylor series expansion near x, y, i.e.,

$$g(x, y) \simeq g(x_0, y_0) + (x - x_0) \frac{\partial g(x_0, y_0)}{\partial x_0} + (y - y_0) \frac{\partial g(x_0, y_0)}{\partial y_0} + \ldots$$

The integral in (7.2) with respect to x, y can now be approximated by

$$\iint g(x, y) f(x, y; t|x_0, y_0; t_0) \mathrm{d}x \mathrm{d}y$$

$$\simeq g(x_0, y_0) + \alpha_x \frac{\partial g(x_0, y_0)}{\partial x_0} + \alpha_y \frac{\partial g(x_0, y_0)}{\partial y_0}$$

$$+ \frac{\alpha_{xx}}{2} \frac{\partial^2 g(x_0, y_0)}{\partial x_0^2} + \ldots. \qquad (7.3)$$

in which the α_x, α_y, α_{xx}, ... are actually functions of x_0, y_0, t, t_0

$$\alpha_x = \iint (x - x_0) f(x, y; t|x_0, y_0; t_0) \mathrm{d}x \mathrm{d}y$$

$$= E\{A(t) - A(t_0)|A(t_0) = x_0, D(t_0) = y_0\}$$

$$\alpha_y = E\{D(t) - D(t_0)|A(t_0) = x_0, D(t_0) = y_0\}$$

$$\alpha_{xx} = \iint (x - x_0)^2 f(x, y; t|x_0, y_0; t_0) \mathrm{d}x \mathrm{d}y$$

$$= E\{[A(t) - A(t_0)]^2|A(t_0) = x_0, D(t_0) = y_0\}, \qquad (7.4)$$

etc., describing the appropriate moments of the changes $A(t) - A(t_0)$, $D(t) - D(t_0)$ given the initial position x_0, y_0.

Finally, substitution of (7.3) into (7.2) gives the expected change in any function of $A(t)$, $D(t)$:

$$E\{g(A(t), D(t))\} - E\{g(A(t_0), D(t_0))\}$$

$$= \iint g(x, y)[f(x, y; t) - f(x, y; t_0)] \, dx \, dy \qquad (7.5)$$

$$= \iint \left\{ \alpha_x \frac{\partial g(x_0, y_0)}{\partial x_0} + \alpha_y \frac{\partial g(x_0, y_0)}{\partial y_0} \right.$$

$$\left. + \frac{\alpha_{xx}}{2} \frac{\partial^2 g(x_0, y_0)}{\partial x_0^2} + \ldots \right\} f(x_0, y_0; t_0) \, dx_0 \, dy_0.$$

So far, we have assumed very little about the process $A(t)$, $D(t)$, only that it changes by a small amount in a short time. Possible complexities in the structure of the process, however, are buried in the coefficients α_x, α_y, . . . which may depend not only upon x_0, y_0, t and t_0 but even upon some past history of the process. In particular, these coefficients are likely to be quite sensitive to x_0, y_0 for x_0, y_0 near any boundary and the effect of a boundary is dictated by the behavior of these coefficients.

Equation (7.5) applies for any smooth function $g(x, y)$ but for certain specific functions it gives obvious identities. For example, if $g(x, y) = 1$, all derivatives of g vanish and (7.5) is merely a statement of the conservation equation

$$\iint f(x, y; t) \, dx \, dy = \iint f(x, y; t_0) \, dx \, dy \text{ for all } t.$$

For $g(x, y) = x$ all derivatives vanish except $\partial g / \partial x$ and (7.5) gives

$$E\{A(t)\} - E\{A(t_0)\} = \iint \alpha_x f(x_0, y_0; t_0) \, dx_0 \, dy_0$$

which is simply a statement of the identity

$$E\{A(t) - A(t_0)\} = E\{E\{A(t) - A(t_0) | A(t_0), D(t_0)\}\}.$$

For $g(x, y) = x^2$, (7.5) gives a basic identity relating unconditional second moments to conditional second moments.

Suppose now that, if there is some boundary, $g(x_0, y_0)$ and/or $f(x_0, y_0; t_0)$ vanish on and near the boundary but α_x, α_y, . . . are differentiable everywhere else. The first term on the right hand side of (7.5) can be integrated by parts to give

$$\int_B \alpha_x g(x_0, y_0) f(x_0, y_0; t) \, dy_0 -$$

$$\iint g(x_0, y_0) \frac{\partial}{\partial x_0} [\alpha_x f(x_0, y_0; t)] \, dx_0 \, dy_0.$$

The first term of this is an integral on the boundary B, which is assumed to vanish. Correspondingly the second derivative terms of (7.5) can be integrated by parts twice with respect to x_0 and/or y_0. If we discard all boundary terms and relabel the integration variables x_0, y_0 as x, y, (7.5) is transformed to

$$\iint g(x, y) \left\{ f(x, y; t) - f(x, y; t_0) + \frac{\partial}{\partial x} [\alpha_x f(x, y; t_0)] \right.$$

$$\left. + \frac{\partial}{\partial y} [\alpha_y f(x, y; t_0)] - \frac{1}{2} \frac{\partial^2}{\partial x^2} [\alpha_{xx} f(x, y; t_0)] + \dots \right\} dx \, dy = 0.$$

Since this integral vanishes for any arbitrary smooth function $g(x, y)$, the coefficient in brackets must vanish at least for values of x, y not too close to any boundary. Thus, by a rather indirect argument, we arrive at an equation describing the change in f itself,

$$f(x, y; t) - f(x, y; t_0)$$

$$\simeq \left[-\frac{\partial}{\partial x} \alpha_x - \frac{\partial}{\partial y} \alpha_y + \frac{1}{2} \frac{\partial^2}{\partial x^2} \alpha_{xx} + \frac{\partial^2}{\partial x \partial y} \alpha_{xy} \right.$$

$$\left. + \frac{1}{2} \frac{\partial^2}{\partial y^2} \alpha_{yy} + \dots \right] f(x, y; t_0) \qquad (7.6)$$

in which all derivatives act on any products to the right, i.e.,

$$\left[\frac{\partial}{\partial x} \alpha_x \right] f(x, y; t_0) \equiv \frac{\partial}{\partial x} \left[\alpha_x f(x, y; t_0) \right].$$

If the α_x, α_y, etc., were all independent of x, y (for x, y not too near a boundary), it would make no difference whether the α_x, etc., appear inside or outside the derivatives. We could have arrived at the same equation by the more direct argument from (7.1).

For $t - t_0$ sufficiently small (but not small compared with an interarrival or service time) and x_0, y_0 not too close to the boundary, we expect that α_x and α_y in (7.4) should be approximately proportional to $(t - t_0)$, i.e.,

$$\beta_x \equiv \frac{\alpha_x}{t - t_0} = E\left\{\frac{A(t) - A(t_0)}{t - t_0}\,\middle|\, A(t_0) = x, D(t_0) = y\right\}$$

$$\beta_y \equiv \frac{\alpha_y}{t - t_0} = E\left\{\frac{D(t) - D(t_0)}{t - t_0}\,\middle|\, A(t_0) = x, D(t_0) = y\right\}$$

should be (nearly) independent of $t - t_0$. The β_x, β_y have the interpretation here of (conditional) arrival and service rates. If we think of $f(x, y; t)$ as a fluid density, β_x, β_y would be interpreted as the average (vector) velocity of the fluid at position x, y and time t_0.

If we write

$$\alpha_{xx} = \text{Var}\left\{A(t) - A(t_0)\,\middle|\, A(t_0) = x, D(t_0) = y\right\} + (\alpha_x)^2$$

then $(\alpha_x)^2$ would be proportional to $(t - t_0)^2$. If α_{xx} is to be proportional to $(t - t_0)$, and therefore of magnitude comparable with α_x, α_y, then α_{xx} must be essentially the conditional variance. For some processes, particularly those which behave like a Brownian motion, α_{xx}, α_{xy}, and α_{yy} will indeed be (nearly) proportional to $(t - t_0)$ so that

$$\beta_{xx} = \frac{\alpha_{xx}}{t - t_0}, \qquad \beta_{xy} = \frac{\alpha_{xy}}{t - t_0}, \qquad \beta_{yy} = \frac{\alpha_{yy}}{(t - t_0)}$$

are also nearly independent of $t - t_0$. In much of the previous discussion we have specifically assumed that

$$\frac{\beta_{xx}}{\beta_x} = \frac{\alpha_{xx}}{\alpha_x} = I_A, \qquad \frac{\beta_{yy}}{\beta_y} = \frac{\alpha_{yy}}{\alpha_y} = I_D, \qquad \text{and } \beta_{xy} = 0.$$

Finally, if we write

$$f(x, y; t) - f(x, y; t_0) \simeq (t - t_0)\frac{\partial f(x, y; t)}{\partial t}$$

then (7.6) becomes

$$\frac{\partial f(x, y; t)}{\partial t} \simeq \left[-\frac{\partial}{\partial x}\beta_x - \frac{\partial}{\partial y}\beta_y + \frac{1}{2}\frac{\partial^2}{\partial x^2}\beta_{xx} + \frac{\partial^2}{\partial x \partial y}\beta_{xy} \right.$$

$$\left. + \frac{1}{2}\frac{\partial^2}{\partial y^2}\beta_{yy} \right] f(x, y; t). \qquad (7.7)$$

This is the standard form of the 'diffusion equation' in two dimensions.

This equation is very similar to equations which arise in the theory of molecular diffusion or heat conduction. In these branches of physics it is customary to introduce the concept of a flux vector, (η_x, η_y),

$$\eta_x(x, y; t) = \left[\beta_x - \frac{1}{2} \frac{\partial}{\partial x} \beta_{xx} - \frac{1}{2} \frac{\partial}{\partial y} \beta_{xy} \right] f(x, y; t)$$

$$\eta_y(x, y; t) = \left[\beta_y - \frac{1}{2} \frac{\partial}{\partial y} \beta_{yy} - \frac{1}{2} \frac{\partial}{\partial x} \beta_{xy} \right] f(x, y; t) \qquad (7.8)$$

so that (7.7) can be written as

$$\frac{\partial f(x, y; t)}{\partial t} + \frac{\partial}{\partial x} \eta_x(x, y; t)$$

$$+ \frac{\partial}{\partial y} \eta_y(x, y; t) = 0. \qquad (7.9)$$

(In various vector notations, the expression $\partial \eta_x / \partial x + \partial \eta_y / \partial y$ may be written as divergence η or $\nabla \cdot \eta$.)

The physical interpretation of the flux vector is that it represents the rate at which mass crosses a line (surface) perpendicular to the direction of the vector per unit length (area) of line (surface). Thus $\eta_x(x, y; t) dy$ and $\eta_y(x, y; t) dx$ are the rates at which mass crosses a line from (x, y) to $(x, y + dy)$ and from (x, y) to $(x + dx, y)$.

If one considers a small rectangle as in Fig. 7.1, the net rate at which mass leaves this rectangle is approximately

$$[\eta_x(x + dx, y; t) - \eta_x(x, y; t)] \, dy + [\eta_y(x, y + dy; t)$$

$$- \eta_y(x, y; t)] \, dx \simeq \left[\frac{\partial \eta_x(x, y; t)}{\partial x} + \frac{\partial n_y(x, y; t)}{\partial y} \right] dx \, dy.$$

The rate of decrease of the mass inside the rectangle is $[-\partial f (x, y; t)/\partial t] \, dx \, dy$. Equation (7.9) then represents a differential form of the equation of conservation of mass; the rate at which the mass in $dx \, dy$ decreases must be equal to the net flow out of the rectangle.

One can also give a physical interpretation to the various terms of η_x, η_y. If a fluid is traveling with a velocity β_x in the x direction, any mass in the rectangle $dx \, dy$ in Fig. 7.1 will have crossed the surface in a time dt such that $\beta_x dt = dx$. Thus a mass $\beta_x f(x, y; t) dy \, dt$ crosses dy in a time dt giving a flux $\beta_x f(x, y; t)$ due to this motion. This explains the first term of (7.8).

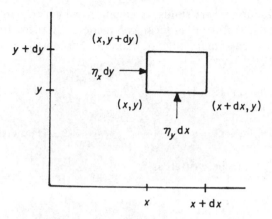

Figure 7.1   *Flow into a small area*

The other flux terms are due to the diffusion. In the theory of heat conduction $f(x, y; t)$ would represent the energy density (temperature) and (7.9) would describe the conservation of energy. The first terms of (7.8) would again represent the flux (of energy) due to transport but the remaining terms describe the heat conduction. If along the left hand face of the rectangle of Fig. 7.1 we had a mass (energy) (density $f(x, y; t)$) and $\beta_x = 0$, there would still be some mass (energy) crossing the left hand face because of dispersion. The rate at which it crosses will be proportional to $f(x, y; t)$ and some diffusion (heat conduction) coefficient β_{xx}. But there will also be a diffusion in the opposite direction across the right hand face proportional to $f(x + dx, y, t)\beta_{xx}$ evaluated at $x + dx$. The net flow across the rectangle will be proportional to the difference in these, i.e., $\partial(f\beta_{xx})/\partial x$.

In the theory of diffusion or heat conduction it is taken as an experimental fact that the mass (energy) flux is proportional to the density (temperature) gradient. In anisotropic crystalline material it is also possible to have different diffusion rates in different directions so that, in general, the flux vector is some linear function of the vector density (temperature) gradient. It is possible to derive the diffusion equation from a model of molecular motion, but the molecular behavior must be experimentally verified. Here we have derived the diffusion equation from some proposed behavior of the process $A(t)$, $D(t)$, but in the final analysis the validity of the equation must be verified experimentally for any situation to which one wishes to apply it.

As applied specifically to the arrival and departure process $A(t)$, $D(t)$, the usual assumptions would be that

$$\beta_x = \lambda(t), \qquad \beta_y = \mu$$
$$\beta_{xx} = I_A \lambda(t), \qquad \beta_{yy} = I_D \mu, \qquad \beta_{xy} = 0, \qquad (7.10)$$

and that these parameters are *a priori* given. There are applications, however, in which β_x could depend upon the queue length $x-y$ because some customers may go away if the queue is too large, or β_y could depend upon the queue length (and/or time); the server may operate more (or less) efficiently under pressure of a rush hour or a long queue. It is, in principle, possible that the coefficients could depend upon the past history of the process (although this would usually make any solution of (7.7) rather difficult). For an infinite-channel server, for example, changes in $A(t)$, $D(t)$ will, in general, depend not only on the past arrival rates but even the actual realizations.

Although as applied to the arrival and departure processes it is usual that $\beta_{xy} = 0$, the same type of equation could be applied to the pair of variables $Q(t)$, $A(t)$ or $Q(t)$, $D(t)$ for which there would be a nonzero covariance rate.

One can also derive a diffusion equation for the arrival and service times $A^{-1}(k), D^{-1}(k)$. It is only necessary to reinterpret the parameter t as the count k.

We have not yet specified the boundary conditions which are, in fact, necessary in order that (7.7) has a unique solution. There are many different types of boundary conditions, however. We will consider them separately as they apply to specific problems.

7.3 Special solutions with no boundaries

Solutions of (7.7) are already known for many cases in which $A(t)$, $D(t)$ stay away from any boundaries.

First of all, if there were no variance terms, $\beta_{xx} = \beta_{yy} = \beta_{xy} = 0$, the equation

$$\frac{\partial f}{\partial t} + \frac{\partial}{\partial x}\beta_x f + \frac{\partial}{\partial y}\beta_y f = 0 \qquad (7.11)$$

would describe a deterministic approximation to the system behavior, but in a rather disguised form. If one were at the position x, y at time t,

β_x, β_y would give not only the expected rate of change of the position but the actual rate of change (i.e., the velocity). For any given starting position $A(t_0)$, $D(t_0)$ at time t_0 there would presumably be a unique location at time t, $A(t)$, $D(t)$ such that

$$\frac{dA(t)}{dt} = \beta_x; \quad \frac{dD(t)}{dt} = \beta_y. \tag{7.12}$$

If β_x, β_y were given functions of $A(t)$, $D(t)$ and t, then (7.12) could be considered as a pair of simultaneous ordinary differential equations for $A(t)$, $D(t)$. It could be solved numerically or graphically, if necessary, but this is not an issue here; we have already discussed solutions of deterministic queueing problems. The main point to observe is that the solution of (7.12) depends upon the initial position $A(t_0)$, $D(t_0)$ at time t_0 and it can therefore be considered as a transformation which for any t maps the point $A(t_0)$, $D(t_0)$ into a point $A(t)$, $D(t)$, i.e., $A(t)$, $D(t)$ is a function of $A(t_0)$, $D(t_0)$.

Equation (7.10) describes the evolution of a probability density $f(x, y; t)$ and the solution of this equation would describe $f(x, y; t)$ in terms of an initial density $f(x_0, y_0; t_0)$. Since the motion is deterministic, any point $A(t_0)$, $D(t_0)$ in an element of area $da_0 = dx_0 dy_0$ maps into a unique point $A(t)$, $D(t)$, and the area element da_0 maps into an element of area da_t. The probability that the system is in an element da_t at time t, $f(x, y; t)da_t$ must be equal to the probability that the system was at time t_0 in any area element da_0 that maps into da_t. If the inverse of the mapping is single valued, then

$$f(x, y; t)da_t = f(x_0, y_0; t_0)da_0. \tag{7.13}$$

The relation (7.13) is simply an application of the standard scheme for determining the joint probability of random variables $A(t)$, $D(t)$ which are known functions of another pair of random variables $A(t_0)$, $D(t_0)$. This is indeed the solution of (7.11).

We will be interested mostly in the special case for which β_x and β_y are independent of x and y (though possibly given functions of t). The solution of (7.11) is then much simpler. Integration of (7.12) gives

$$A(t) - A(t_0) = \int_{t_0}^{t} \beta_x(t') dt'$$

$$D(t) - D(t_0) = \int_{t_0}^{t} \beta_y(t') dt'$$

which for any t describes simply a uniform translation of all points $A(t_0)$, $D(t_0)$ into $A(t)$, $D(t)$. The probability fluid moves like a 'rigid body' with $da_t = da_0$. The explicit solution of (7.11), which is now equivalent to the equation

$$\frac{\partial f}{\partial t} + \beta_x \frac{\partial f}{\partial x} + \beta_y \frac{\partial f}{\partial y} = 0,$$

is

$$f(x, y; t) = f\left(x - \int_{t_0}^{t} \beta_x(t')dt', y - \int_{t_0}^{t} \beta_y(t')dt'; t_0 \right). \quad (7.14)$$

This describes the density at time t in terms of the density at time t_0.

The general solution of (7.7) with the coefficients β_x, β_y, β_{xx}, given functions x, y and t can be very complex, but there are simple solutions if all coefficients are functions of t only but independent of x, y (and there are no boundaries). We know the solution in this case because (7.7) must describe the evolution of the distribution $A(t)$, $D(t)$ when changes in $A(t)$, $D(t)$ represent a vector Brownian motion. The distribution of $A(t) - A(t_0)$, $D(t) - D(t_0)$ must be bivariate normal with

$$E\{A(t) - A(t_0)\} = \int_{t_0}^{t} \beta_x(t')dt'$$

$$E\{D(t) - D(t_0)\} = \int_{t_0}^{t} \beta_y(t')dt'$$

$$\text{Var}\{A(t) - A(t_0)\} = \int_{t_0}^{t} \beta_{xx}(t')dt'$$

$$\text{Var}\{D(t) - D(t_0)\} = \int_{t_0}^{t} \beta_{yy}(t')dt'$$

$$\text{Cov}\{A(t) - A(t_0), D(t) - D(t_0)\} = \int_{t_0}^{t} \beta_{xy}(t')dt'.$$

The complete solution for $f(x, y; t)$ is obtained by substituting this conditional distribution into (7.1) and integrating over the initial distribution. One can indeed verify that this is a solution of (7.7) for any initial distribution by substituting the proposed solution into (7.7).

For the case in which the coefficients have the form (7.10), the solution of (7.7) is

$$f(x, y; t) = \int\!\!\int f(x_0, y_0; t_0) \left[2\pi I_A I_D \mu (t - t_0) \int_{t_0}^{t} \lambda(t')dt' \right]^{-1/2}$$

$$\times \exp \left\{ - \frac{\left[x - x_0 - \int_{t_0}^{t} \lambda(t')dt' \right]^2}{2 I_A \int_{t_0}^{t} \lambda(t')dt'} \right.$$

$$\left. - \frac{(y - y_0 - \mu t)^2}{2 I_D \mu (t - t_0)} \right\} dx_0 dy_0. \tag{7.15}$$

Even if the variance rates depend upon x, y but the first moments are independent of x, y, one can simplify (7.7) by going to a moving coordinate system. If we let

$$x'(x, t) = x - \int_{t_0}^{t} \beta_x(t')dt', \quad y'(x, t) = y - \int_{t_0}^{t} \beta_y(t')dt'$$

and

$$f'(x', y'; t) = f(x(x', t), y(y', t); t)$$

then x', y' would represent the spatial coordinates relative to an origin moving with the fluid (coordinates relative to the center of mass) and $f'(x', y'; t)$ would be the probability density relative to the new coordinates. In the new coordinates, $f'(x', y'; t)$ will satisfy an equation of the form

$$\frac{\partial f'(x', y'; t)}{\partial t} = \frac{1}{2} \frac{\partial^2}{\partial x'^2} \beta'_{xx} f' + \frac{\partial^2}{\partial x' \partial y'} \beta'_{xy} f' + \frac{1}{2} \frac{\partial^2}{\partial y'^2} \beta'_{yy} f' \cdots$$

with no first derivative terms. There is even a further generalization of this scheme when the β_x and β_y depend upon x, y. It would exploit the fact that (7.13) is a solution of (7.11). Unfortunately, these transformations to a moving coordinate system are not very helpful in evaluating solutions with boundary conditions because they will transform a stationary boundary into a moving boundary.

These solutions obtained without any boundaries are useful in describing certain dynamical behavior because they still apply to any part of the distribution which is not near a boundary. Equation (7.7) satisfies the basic linearity property that if $f_1(x, y; t)$ and $f_2(x, y; t)$ are solutions, so is $f_1 + f_2$. One of these solutions can be chosen so that it stays away from any boundary at least for some finite time period.

7.4. Marginal distributions and boundary conditions

Equation (7.7) describes the evolution of the joint probability density for any pair of random variables which have the appropriate local cause and effect behavior. If these conditions apply for random variables $A(t)$, $D(t)$, they will also apply for any linear functions of the $A(t)$, $D(t)$, in particular for the pair of variables $A(t)$, $Q(t) = A(t) - D(t)$ or $D(t)$, $Q(t)$.

Equations for the joint probability density of $A(t)$, $Q(t)$, or $D(t)$, $Q(t)$ can be derived directly by specifying the appropriate coefficients, or by specifying the coefficients for the $A(t)$, $D(t)$ process and making a transformation of variables. If we let

$$f_{AQ}(x, l; t) = f_{AD}(x, y; t)$$
$$\text{and} \quad f_{DQ}(y, l; t) = f_{AD}(x, y; t) \quad \text{with} \quad l = x - y, \tag{7.16}$$

then f_{AQ} and f_{DQ} would represent the joint probability densities for $A(t)$, $Q(t)$ and $D(t)$, $Q(t)$, respectively (since the Jacobians of the transformations are both equal to 1). Derivatives of any function of x, y, t can be converted into derivatives of the corresponding function of x, l, t or y, l, t through the relations

$$\frac{\partial f_{AD}(x, y; t)}{\partial x} = \left(\frac{\partial}{\partial x} + \frac{\partial}{\partial l} \right) f_{AQ}(x, l; t) = \frac{\partial}{\partial l} f_{DQ}(y, l; t)$$

$$\frac{\partial f_{AD}(x, y; t)}{\partial y} = - \frac{\partial}{\partial l} f_{AQ}(x, l; t) \qquad = \left(\frac{\partial}{\partial y} - \frac{\partial}{\partial l} \right) f_{DQ}(y, l; t)$$

$$\frac{\partial}{\partial t} f_{AD}(x, y; t) = \frac{\partial}{\partial t} f_{AQ}(x, l; t) \qquad = \frac{\partial}{\partial t} f_{DQ}(x, l; t).$$

Substitution of these differential identities into (7.7) will convert the equation for $f_{AD}(x, y; t)$ into corresponding equations for $f_{AQ}(x, l; t)$ or $f_{DQ}(y, l; t)$. If, as is usually the case, $\beta_{xy} = 0$ for the process $A(t)$, $D(t)$ then f_{AQ} and f_{DQ} satisfy the equations,

$$\frac{\partial f_{AQ}}{\partial t} = \left[- \frac{\partial}{\partial x} \beta_x - \frac{\partial}{\partial l} (\beta_x - \beta_y) \right.$$
$$\left. + \frac{1}{2} \frac{\partial^2}{\partial x^2} \beta_{xx} + \frac{\partial^2}{\partial x \partial l} \beta_{xx} + \frac{1}{2} \frac{\partial^2}{\partial l^2} (\beta_{xx} + \beta_{yy}) \right] f_{AQ} \tag{7.17}$$

$$\frac{\partial f_{DQ}}{\partial t} = \left[- \frac{\partial}{\partial y} \beta_y - \frac{\partial}{\partial l} (\beta_x - \beta_y) \right.$$
$$\left. + \frac{1}{2} \frac{\partial}{\partial y^2} \beta_{yy} - \frac{\partial^2}{\partial y \partial l} \beta_{yy} + \frac{1}{2} \frac{\partial^2}{\partial l^2} (\beta_{xx} + \beta_{yy}) \right] f_{DQ}$$

in which the coefficients β_x, β_y, etc., are coefficients of the process $A(t)$, $D(t)$ expressed as functions of x and l or y and l respectively. These equations merely demonstrate the fact that $Q(t)$ has an expected growth rate $\beta_x - \beta_y$ but that $A(t)$, $Q(t)$ and $D(t)$, $Q(t)$ have covariance rates β_{xx} and $-\beta_{yy}$, respectively (if $\beta_{xy} = 0$).

If there are any boundaries, they are likely to affect only $Q(t)$; $Q(t)$ must satisfy the condition $Q(t) > 0$ and possibly a condition $Q(t) < c$ if there is only a finite storage space c (which could be time-dependent). If there are no restrictions on x or y, we can integrate both sides of (7.17) with respect to x or y for fixed value of l from $-\infty$ to $+\infty$. Any boundary terms at $x, y = \pm \infty$ are assumed to vanish, so

$$\frac{\partial}{\partial t} \int_{-\infty}^{+\infty} f_{AQ}(x, l; t) \mathrm{d}x$$

$$= -\frac{\partial}{\partial l} \int_{-\infty}^{+\infty} (\beta_x - \beta_y) f_{AQ}(x, l; t) \mathrm{d}x$$

$$+ \frac{1}{2} \frac{\partial^2}{\partial l^2} \int_{-\infty}^{-\infty} (\beta_{xx} + \beta_{yy}) f_{AQ}(x, l; t) \mathrm{d}x, \quad (7.18)$$

and a corresponding equation for f_{AD}.

The left hand side of (7.18) can be interpreted as the time derivative of the marginal probability density of $Q(t)$,

$$f_Q(x; t) = \int_{-\infty}^{+\infty} f_{AQ}(x, l; t) \mathrm{d}x.$$

The two integrals on the right hand side can be interpreted as rates at which $E\{Q(t)\}$ and $\mathrm{Var}\{Q(t)\}$ increase for $Q(t) = l$. If the coefficients $\beta_x, \ldots$ depend upon x, the values of these integrals will depend upon the distribution of $A(t)$ for given l, but if these coefficients are independent of x (but still possibly functions of l and t) then (7.18) reduces to an equation for $f_Q(l; t)$ alone

$$\frac{\partial}{\partial t} f_Q(l; t) = -\frac{\partial}{\partial l} (\beta_x - \beta_y) f_Q(l; t) + \frac{1}{2} \frac{\partial^2}{\partial l^2} (\beta_{xx} + \beta_{yy}) f_Q(l; t),$$

$$(7.19)$$

i.e., $f_Q(l; t)$ itself satisfies a one-dimensional diffusion equation. Integration of the equation for $f_{DQ}(y, l; t)$ with respect to y leads to the same equation under the condition that the coefficients are functions only of l and t.

Integration of (7.19) with respect to l from 0 to c gives

$$\frac{\partial}{\partial t} \int_0^c f_Q(l; t) \, dl = \left[-(\beta_x - \beta_y) f_Q(l; t) \right.$$
$$\left. + \frac{1}{2} \frac{\partial}{\partial l} (\beta_{xx} + \beta_{yy}) f_Q(l; t) \right]_{l=0}^c \qquad (7.20)$$

The left hand side of (7.20) represents the total probability for $0 < Q(t) < c$. If any probability which hits a boundary stays only for a short time (of the order of an interarrival or service time) so that $P\{Q(t) = 0\}$ or $P\{Q(t) = c\}$ can be neglected (i.e., the continuum approximation applies right to the boundary), then the left hand side of (7.20) is $(\partial/\partial t)1 = 0$. The terms on the right hand side evaluated at 0 and c, represent the flow of probability out of the region $(0, c)$ through the boundaries. If it were possible for the mass which reaches $l = 0$ to jump suddenly to $l = c$, it would be possible for the right hand side of (7.20) to vanish without the flow at each boundary vanishing separately. Clearly, this is not the type of condition in which we are interested here.

We conclude that if any probability reaching a boundary cannot stay and cannot jump then

$$-(\beta_x - \beta_y) f_Q(l, t) + \frac{1}{2} \frac{\partial}{\partial l} (\beta_{xx} + \beta_{yy}) f_Q(l, t) = 0$$

$$\text{at } x = 0 \quad \text{and} \quad x = c \quad \text{for all } t. \qquad (7.21)$$

In analyzing the evolution of the queue distribution, it is often more convenient to deal with distribution function than the probability density. If

$$F_Q(l; t) = \int_0^l f_Q(l'; t) \, dl'$$

and $f_Q(l; t)$ satisfies (7.21) at least for $x = 0$, then an integration of (7.19) yields

$$\frac{\partial}{\partial t} F_Q(l; t) = -(\beta_x - \beta_y) \frac{\partial}{\partial l} F_Q(l; t)$$

$$+ \frac{1}{2} \frac{\partial}{\partial l} (\beta_{xx} + \beta_{yy}) \frac{\partial F_Q(l; t)}{\partial l} \qquad (7.22)$$

with the boundary condition

$$F_Q(0; t) = 0 \quad \text{for all } t.$$

The second condition, that probability cannot leave through $x = c$, obviously implies that

$$F_Q(c; t) = 1 \quad \text{for all } t.$$

Suppose, on the other hand, that $x = 0$ and/or $x = c$ were 'absorbing boundaries', any probability that hits the boundary stays there. It is the nature of the Brownian motion that for short times the motion is dominated by the fluctuations, which in a time interval dt are proportional to $(dt)^{1/2}$. If any probability mass comes close enough to the boundary, it is almost certain to hit the boundary within a short time; it cannot escape. If there is absorption at $x = c$, for example, then the appropriate boundary condition is

$$f_Q(c; t) = 0 \quad \text{or} \quad \partial F_Q(l; t)/\partial l = 0 \quad \text{at} \quad l = c \qquad (7.23)$$

for (7.21) or (7.22) respectively.

With this boundary condition (7.20) now has the interpretation that it describes the rate at which probability is lost to the boundary. The solution of (7.22) represents the probability that $Q(t) < l$ and that $Q(t') < c$ for all t', $0 < t' < t$. Correspondingly $F_Q(c; t)$ is the probability that $Q(t') < c$ for $0 < t' < t$.

The solution $F_Q(c; t)$ of (7.22) with absorption at c is of importance in many applications because

$$F_Q(c; t) = P\{Q(t') < c, \qquad 0 < t' \leq t\}$$

$$= P\{\max_{t'} Q(t') < c, \ 0 < t' \leq t\}. \qquad (7.24)$$

Thus, considered as a function of c for fixed t, it is the distribution function for $\max Q(t')$; the probability that one will not need more storage than c during the time t. On the other hand, if T_c represents the time at which $Q(t)$ reaches the boundary for the first time

$$F_Q(c; t) = P\{T_c > t\}. \qquad (7.25)$$

Thus, considered as a function of t for fixed c, $1 - F_Q(c, t)$ is the distribution function for the 'first passage time', T_c.

Returning to the joint distributions, suppose that we now integrate

(7.17) with respect to l from 0 to c for fixed x or y so as to obtain

$$\frac{\partial}{\partial t}\int_0^c f_{AQ}(x, l; t)dl = -\frac{\partial}{\partial x}\int_0^c \beta_x f_{AQ}(x, l; t)dl$$

$$+ \frac{1}{2}\frac{\partial^2}{\partial x^2}\int_0^c \beta_{xx}f_{AQ}(x, l; t)dl$$

$$+ \left[-(\beta_x - \beta_y) + \frac{\partial}{\partial x}\beta_{xx} + \frac{1}{2}\frac{\partial}{\partial l}(\beta_{xx}\right.$$

$$\left. + \beta_{yy})\right] f_{AQ}(x, l; t)\Bigg|_{l=0}^{c}$$

$$(7.26)$$

and a corresponding equation for f_{DQ}. The left hand side of (7.26) represents the time derivative of the marginal probability density of $A(t)$

$$f_A(x; t) = \int_0^c f_{AQ}(x, l; t)\,dl$$

provided there is a negligible probability on the boundaries $Q(t) = 0$ or c. The two integrals on the right hand side describe the rates of increase of the first and second moments of $A(t)$ for $A(t) = x$. Although these integrals cannot generally be evaluated in terms of $f_A(x; t)$ if the coefficients β_x, β_{xx} are functions of l, their meaning is clear. They would describe the change of $f_A(x; t)$ if there were no boundaries. The terms evaluated at $l = 0$ or c must therefore represent something which happens at (or near) the boundaries that affects the arrival process.

If there is a finite storage c and the storage becomes full something must indeed be done about any arrivals who have no place to go. If the conservation equation is to remain valid under all conditions, $A(t)$ must be interpreted as the arrivals who actually enter the system. If anyone who might arrive when the storage is full is rejected, the term of (7.26) for $l = c$ must represent the rate at which f_A changes because some customers who try to arrive with mean and variance rates β_x, β_{xx} when $Q(t) = l$ are turned away.

The boundary term at $x = 0$ describes a corresponding phenomenon that could, in principle, occur if the queue vanishes. One could imagine a process $A(t)$, $Q(t)$ satisfying (7.17) for which the arrival process creates some extra customers when the queue vanishes so as to keep the server busy. The type of process we have in mind, however, is one for which a vanishing queue has no effect upon the arrival process. For this to be true, it is necessary to impose a boundary condition on $f_{AQ}(x; t)$.

$$\left[-(\beta_x - \beta_y) + \frac{\partial}{\partial x}\beta_{xx} + \frac{1}{2}\frac{\partial}{\partial l}(\beta_{xx} + \beta_{yy}) \right] f_{AQ}(x, l; t) = 0 \quad \text{at } l = 0.$$

(7.27)

Since f_{AQ}, f_{DQ}, and f_{AD} are related through (7.16), a boundary condition on f_{AQ} at $l = 0$ induces a boundary condition on f_{DQ} at $l = 0$ and on f_{AD} at $x = y$, namely

$$\left[-(\beta_x - \beta_y) + \frac{1}{2}\frac{\partial}{\partial y}(\beta_{xx} - \beta_{yy}) + \frac{1}{2}\frac{\partial}{\partial l}(\beta_{xx} + \beta_{yy}) \right] f_{DQ}(y, l; t)$$

$$= 0 \quad \text{at } l = 0$$

(7.28)

and

$$\left[-(\beta_x - \beta_y) + \frac{\partial}{\partial x}\beta_{xx} + \frac{1}{2}\frac{\partial}{\partial y}(\beta_{xx} - \beta_{yy}) \right] f_{AD}(x, y; t)$$

$$= 0 \quad \text{at } x = y.$$

(7.29)

These boundary conditions are not the same as those which usually arise in particle diffusion or heat conduction. In the latter situations, it is usually the normal component of the flux that vanishes. Here the boundary forces probability mass approaching the boundary to 'slide' along the boundary as if to exert a viscous drag slowing down the $D(t)$ process when the server runs out of customers.

We still must obtain some boundary conditions at $l = c$. When a storage becomes full, one could imagine a process for which the customers who could not enter the storage area are passed directly through the server rather than being turned away, thereby giving an extra surge to the $D(t)$ process when $Q(t)$ hits c. We are primarily concerned here, however, with a process for which a full storage has no immediate effect on the $D(t)$. The server serves only the customers who have entered the storage.

If we integrate (7.17) for f_{DQ} with respect to l we conclude that f_{DQ} must satisfy the boundary condition.

$$\left[-(\beta_x - \beta_y) - \frac{\partial}{\partial y}\beta_{yy} + \frac{1}{2}\frac{\partial}{\partial l}(\beta_{xx} + \beta_{yy}) \right] f_{DQ}(y, l; t)$$

$$= 0 \quad \text{for } l = c$$

(7.30)

if $Q(t) = c$ is to have no effect upon the departure process. This induces the boundary conditions on f_{AQ} and f_{AD}.

$$\left[-(\beta_x - \beta_y) + \frac{1}{2}\frac{\partial}{\partial x}(\beta_{xx} - \beta_{yy}) + \frac{1}{2}\frac{\partial}{\partial l}(\beta_{xx} + \beta_{yy}) \right] f_{AQ}(x, l; t)$$

$$= 0 \quad \text{for } l = c \qquad (7.31)$$

$$\left[-(\beta_x - \beta_y) - \frac{\partial}{\partial y}\beta_{yy} + \frac{1}{2}\frac{\partial}{\partial x}(\beta_{xx} - \beta_{yy}) \right] f_{AD}(x, y; t)$$

$$= 0 \quad \text{for } x = y. \qquad (7.32)$$

If f_{AQ} satisfies the boundary condition (7.27) at $l = 0$ and (7.31) at $l = c$, the boundary term of (7.26) can be written in the form

$$\frac{1}{2}\frac{\partial}{\partial x}(\beta_{xx} + \beta_{yy})f_{AQ}(x, l; t)\Big|_{l=c}.$$

If, in addition, the coefficients β_x and β_{xx} are independent of l, i.e., the arrival process does not depend upon the queue length except if $Q(t) = c$, (7.26) simplifies to

$$\frac{\partial}{\partial t}f_A(x; t) = \left[-\frac{\partial}{\partial x}\beta_x + \frac{1}{2}\frac{\partial^2}{\partial x^2}\beta_{xx} \right] f_A(x; t)$$

$$+ \frac{1}{2}\frac{\partial}{\partial x}(\beta_{xx} + \beta_{yy})f_{AQ}(x, l; t)\Big|_{l=c}. \qquad (7.33)$$

For $c = \infty$ there would be no boundary term and $f_A(x; t)$ would itself satisfy a diffusion equation with no boundary conditions. In particular, if $\beta_x = \lambda(t)$, $\beta_{xx} = I_A\lambda(t)$ are also independent of x, the solution of this equation starting from a position $A(t_0) = x_0$ would obviously be a normal distribution with

$$E\{A(t)\} = x_0 + \int_{t_0}^{t} \lambda(t')dt', \quad \text{Var}\{A(t)\} = I_A \int_{t_0}^{t} \lambda(t')dt'.$$

For $c < \infty$, (7.33) shows that the behavior of the arrival process depends upon the probability density of $Q(t)$ at the boundary (some arrivals are turned away if $Q(t) = c$) and the variance coefficients.

If we multiply both sides of (7.33) by x and integrate from $-\infty$ to $+\infty$, we obtain

$$\frac{d}{dt}E\{A(t)\} = \beta_x - \frac{1}{2}(\beta_{xx} + \beta_{yy})f_Q(l; t)\Big|_{l=c}. \qquad (7.34)$$

The second term on the right is the average rate at which customers are lost. Similarly,

$$\frac{d}{dt} \text{Var}\{A(t)\} = \beta_{xx} - (\beta_{xx} + \beta_{yy})f_Q(l, t)\Big|_{l=c}$$

$$\times [E\{A(t)|Q(t) = c\} - E\{A(t)\}]. \quad (7.35)$$

There are corresponding relations for the departure process. If the departure rates β_y and β_{yy} are independent of the queue length for $Q(t) > 0$, then the counterparts of (7.33), (7.34), and (7.35) are

$$\frac{\partial}{\partial t} f_D(y; t) = \left[-\frac{\partial}{\partial y}\beta_y + \frac{1}{2}\frac{\partial^2}{\partial y^2}\beta_{yy} \right] f_D(y; t)$$

$$+ \frac{1}{2}\frac{\partial}{\partial y}(\beta_{xx} + \beta_{yy})f_Q(y, l; t)\Big|_{l=0}, \quad (7.36)$$

$$\frac{d}{dt}E\{D(t)\} = \beta_y - \frac{1}{2}(\beta_{xx} + \beta_{yy})f_Q(l; t)\Big|_{l=0}, \quad (7.37)$$

and

$$\frac{d}{dt}\text{Var}\{D(t)\} = \beta_{yy} - (\beta_{xx} + \beta_{yy})f_Q(l; t)\Big|_{l=0}$$

$$\times [E\{D(t)|Q(t) = 0\} - E\{D(t)\}]. \quad (7.38)$$

The second term of (7.37), in particular, describes the average loss of departure rate caused by a lack of customers to serve when $Q(t) = 0$

CHAPTER 8

Diffusion approximation for equilibrium and transient queue behavior

8.1 Equilibrium distributions

The difficulty in solving diffusion equations generally increases with the number of dimensions. Although the evolution of the joint density $f(x, y; t)$ for both the arrivals and the departures would give a fairly complete description of what is happening and why, it is much easier to analyze the behavior of only $Q(t)$. We have seen that if the mean and variance rates for $A(t)$ and $D(t)$ depend only upon the queue length and time, then $f_Q(l; t)$ itself satisfies a diffusion equation (7.19). If there is a negligible probability that $Q(t)$ is on a boundary, then $f_Q(l; t)$ also satisfies the boundary conditions (7.21) and $F_Q(l; t)$ satisfies (7.22).

If the coefficients β_x, β_y, β_{xx} and β_{yy} are independent of t and functions only of l, we would expect an equilibrium distribution to exist such that

$$f_Q(l; t) \to f_Q(l) \quad \text{for} \quad t \to \infty$$

for any initial queue length at $t = t_0$. If so, $f_Q(l)$ must satisfy the ordinary differential equation

$$-\frac{\mathrm{d}}{\mathrm{d}l}(\beta_x - \beta_y)f_Q(l) + \frac{1}{2}\frac{\mathrm{d}^2}{\mathrm{d}l^2}(\beta_{xx} + \beta_{yy})f_Q(l) = 0.$$

This equation can be integrated with respect to l to give

$$-(\beta_x - \beta_y)f_Q(l) + \frac{1}{2}\frac{\mathrm{d}}{\mathrm{d}l}(\beta_{xx} + \beta_{yy})f_Q(l) = C$$

for some integration constant C, but the boundary conditions at both

$x = 0$ and $x = c$ imply that $C = 0$. The second boundary condition is redundant. This first order equation can now be integrated to give

$$f_Q(l) = \frac{C'}{(\beta_{xx}+\beta_{yy})} \exp\left[2\int_0^l \frac{(\beta_x - \beta_y)dl'}{(\beta_{xx}+\beta_{yy})} \right]$$

for some other integration constant C'.

Since both the differential equation and the boundary conditions are linear in f, any multiple of a solution of these equations is also a solution. The constant C' must be determined from the normalization

$$\int_0^c f_Q(l')dl' = 1.$$

Thus the equilibrium solution is

$$f_Q(l) = \frac{\dfrac{1}{(\beta_{xx}+\beta_{yy})} \exp\left[2\int_0^l \dfrac{(\beta_x - \beta_y)dl'}{(\beta_{xx}+\beta_{yy})} \right]}{\displaystyle\int_0^c \dfrac{1}{(\beta_{xx}+\beta_{yy})} \exp\left[2\int_0^{l'} \dfrac{(\beta_x - \beta_y)dl''}{(\beta_{xx}+\beta_{yy})} \right]dl'}. \tag{8.1}$$

We are particularly interested in the special case in which $\beta_x = \lambda$, $\beta_y = \mu$, $\beta_{xx} = I_A\lambda$, and $\beta_{yy} = I_D\mu$, which are all independent of l. In this case (8.1) gives

$$f_Q(l) = \frac{(2/L_0)\exp(-2l/L_0)}{1-\exp(-2c/L_0)} \quad \text{for} \quad \mu \neq \lambda \tag{8.2}$$

$$= \frac{1}{c} \quad \text{for} \quad \mu = \lambda$$

with

$$L_0 = \frac{I_D + I_A\rho}{1-\rho} \tag{8.3}$$

as defined previously in (5.5). Also

$$F_Q(l) = \frac{1-\exp(-2l/L_0)}{1-\exp(-2c/L_0)} \quad \text{for} \quad \mu \neq \lambda$$

$$= l/c \quad \text{for} \quad \mu = \lambda.$$

All formulas apply only for $0 < l < c$.

Since (7.34) and (7.37) involve only the queue length distribution, it follows that there must be an equilibrium rate at which customers pass through the system.

$$\frac{dE\{A(t)\}}{dt} = \lambda - \frac{(\mu - \lambda)}{1 - \exp(-2c/L_0)}$$

$$= \frac{dE\{D(t)\}}{dt} = \mu - \frac{(\mu - \lambda)\exp(-2c/L_0)}{1 - \exp(-2c/L_0)}$$

$$= \frac{\lambda + \mu}{2} - \frac{(\mu - \lambda)}{2}\text{ctnh}(c/L_0) \quad \text{for} \quad \mu \neq \lambda$$

$$= \mu - \frac{(I_D + I_A)}{2c}\mu \qquad \text{for} \quad \mu = \lambda. \qquad (8.4)$$

The diffusion equation was derived under the hypothesis that the discrete nature of the queue could be neglected and that the coefficients $\beta_x, \ldots$ did not depend upon past history, i.e., the arrival and departure processes behaved approximately like a Brownian motion. These are similar hypotheses to those discussed in Chapter 5 where we argued on dimensional grounds that the typical queue length for $c = \infty$ should be comparable with L_0. We also showed in Chapter 5 that the queue distribution for the $M/M/1/c$ system was a truncated geometric distribution.

The truncated exponential distribution obtained here is, of course, the continuum analogue of a geometric distribution. What we have shown so far is that the exponential (geometric) form of the queue distribution is relatively insensitive to the detailed properties of the arrivals and departures if $c \gg 1$ and $|L_0| \gg 1$. The scale of the distribution L_0, however, depends upon the variance rates I_A and I_D.

Since these equilibrium approximations are valid only for $|L_0| \gg 1$, i.e., $|1 - \rho| \ll 1$ and $c \gg 1$, one should not expect them to give any more than the leading terms in the expansion of any stochastic property in powers of $1 - \rho$, or to measure queues more accurately than the nearest integer. For example, if $c = \infty$ and $\rho < 1$, the above approximations give

$$E\{Q_s\} = L_0/2 = \frac{I_D + I_A\rho}{2(1 - \rho)} = \frac{I_D + I_A}{2(1 - \rho)} - \frac{I_A}{2} \qquad (8.5)$$

whereas the exact formula for the $M/M/1$ system ($I_A = I_D = 1$) given by (5.18) is

$$E\{Q_s\} = \frac{\rho}{1 - \rho} = \frac{1}{1 - \rho} - 1.$$

The approximate formula overestimates $E\{Q_s\}$ by 1/2. Much of this error can probably be traced to the fact that in the diffusion approximation the properties of $D(t)$ are obtained under the hypothesis that the server is busy most of the time (an overestimation of $\text{Var}\{D(t)\}$), but it is difficult to justify trying to make corrections to a continuum approximation for less than one customer.

For $|1 - \rho| \ll 1$, the leading term in an expansion of (8.4) in powers of $1 - \rho$ would be $dE\{A(t)\}/dt = \lambda$ or μ but this approximation could have been derived from a deterministic fluid approximation. The diffusion approximation gives a valid estimate of the next term, the blocking effect. If we compare (8.4) with (5.26) we see, for example, that for $\rho = 1$, (5.26) gives

$$\frac{dE\{A(t)\}}{dt} = \mu - \frac{\mu}{c + 1}$$

whereas (8.4) gives $\mu - \mu/c$ for $I_A = I_D = 1$. Again the diffusion approximations slightly overestimate the effect of fluctuation on blocking but only by a factor of $(c + 1)/c$. We could not expect the diffusion approximation to distinguish between a storage of c or $c + 1$. The $c + 1$ could be interpreted as the storage of a system which includes a zeroth input server; in fact, this is a proper interpretation for the $M/M/1/c$ system.

We do not necessarily expect that the diffusion approximation will describe the behavior of other systems with I_A and/or $I_D \neq 1$ as accurately as it does the $M/M/1$ system. The error in $E\{Q_s\}$, for example, for $|1 - \rho| \ll 1$ should be of order 1 relative to $|1 - \rho|$ if the arrival and departure processes define an I_A and I_D, i.e., $\text{Var}\{A(t)\}$ and $\text{Var}\{D(t)\}$ are (nearly) proportional to t. If, for example, customers arrive in batches, the diffusion approximation could not be expected to estimate the average queue to an accuracy less than a typical batch size.

It should be emphasized again that the goal here is not necessarily to achieve high accuracy. If one knew the parameters of any proposed system one could always do a simulation to evaluate the behavior as accurately as one wished. Our main goal is to describe, in as simple a way as possible, approximately how the system behavior depends upon any relevant parameters so that one can decide what parameter values are worthy of consideration in designing or modifying some service system. For example, one can see immediately from the above formulas approximately how a reduction in the arrival and service variance might affect the equilibrium queue or how an increase in storage c might affect the flow through the system. Simulation is a powerful tool for validation but a clumsy tool for searching. Complicated formulas are of questionable value for either purpose.

Although most of our attention has been directed toward the single-channel server, the diffusion approximation can also be applied to the multiple-channel server. If there are only a few channels and $1 - \rho \ll 1$, all servers would be busy most of the time and the system would behave like a single-channel server with service rate $m\mu$. For m servers having identical and independent service times the departure process while the system is busy is the sum of m independent processes. The Var $\{D(t)\}/E\{D(t)\} \simeq I_D$ for the sum is the same as for the individual servers (see problem 4.10).

A common question regarding the multiple-server system is the following. If one had m independent single-channel queueing systems with identical servers and the kth server served an arrival process with rate λ_k, variance rate $I_{Ak}\lambda_k$, what advantage would result if one combined the servers into an m-channel system serving a common queue with arrival rate $\Sigma \lambda_k$? If the λ_k and I_{Ak} were different, it would first of all be advantageous generally to redistribute the traffic to the single servers so that they had nearly the same λ_i. We will, therefore, suppose that λ_i and I_{Ai} are the same for all servers.

If the component arrival processes were independent, the superposition of them would have the same I_A. The combined system would also have the same I_D and ρ as the separate systems. In the diffusion approximation the queue distribution depends only upon I_A, I_D, and ρ (but is independent λ itself); consequently, the combined system would have approximately the same queue distribution as each of the separate systems. The conclusion is that the total number of customers in the system would drop by approximately a factor of m if the independent single-channel systems were combined into a single m-channel system. The mean waiting time would also drop by a factor of m, because the combined system is served at a rate $m\mu$.

It is issues such as these that motivate airlines to pool their gate positions at an airport or a telephone company to use large capacity trunk lines. The situation at a toll plaza, grocery store cashier, or bank, however, is quite different. Even though there may be separate queues behind each server, a newly arriving customer will usually join the shortest queue (the arrivals to the different servers and the queues are not independent). The queue for the latter system is almost the same as for a multiple-channel server provided each server will be busy whenever there is anyone to serve.

The diffusion approximation can also be applied when $m \gg 1$, even to the infinite-channel server, but only under rather special conditions. If the number of customers in the system is large compared with 1, one can use a continuum approximation. It is also true that the number of customers in the system will not change very much in a short time but the β_y and β_{yy} (for the departures) will generally depend upon when the customers in the system arrived. One can use the diffusion equation, however, if the service times are exponentially distributed and the arrivals behave like a Brownian motion with variance rate $I_A \lambda$ (not necessarily as a Poisson process).

For a system with m identical channels $m \gg 1$, exponential service time, and storage c

$$\beta_x = \lambda, \quad \beta_{xx} = I_A \lambda \quad \text{and} \quad \beta_y = \beta_{yy} = \begin{cases} \mu l & 0 \leq l \leq m \\ \mu m & m \leq l \leq c. \end{cases}$$

Substitution of these parameters into (8.1) gives

$$f_Q(l) = \begin{cases} C\left[1 + \dfrac{l}{m\rho I_A}\right]^{2m\rho(1+I_A)-1} e^{-2l} & \text{for} \quad 0 \leq l \leq m \\[4mm] f_Q(m)\exp\left[-\dfrac{2(1-\rho)l}{I_A\rho + 1}\right] & \text{for} \quad m \leq l \leq c \end{cases} \qquad (8.6)$$

with C a normalization constant which can be expressed as a rather clumsy integral.

The exponential distribution for $m \leq l \leq c$ has the same form as (8.2) with $I_D = 1$, in agreement also with the geometric tail associated with M/M/m system. The formula for $0 \leq l \leq m$, however, is not immediately recognizable as an approximation to the known Poisson shaped distribution for the M/M/m system given in (5.20). Actually, it is a more accurate approximation than the normal approximation given in (5.22).

If we set the derivative of $f_Q(l)$ equal to zero, we see that $f_Q(l)$ has a maximum at

$$l_0 = m\rho - 1/2 = \lambda/\mu - 1/2.$$

Since λ/μ is the expected number of busy servers (for $c = \infty$) the continuum approximation would be meaningful only for $\lambda/\mu \gg 1$. We would have obtained the normal approximation to $f_Q(l)$ if we had approximated the variance coefficient β_{yy} by its value at $l = \lambda/\mu$, i.e.,

$\beta_{xx} \simeq \lambda$ so that in (8.1)

$$\frac{\beta_x - \beta_y}{\beta_{xx} + \beta_{yy}} \simeq \frac{\lambda - \mu l}{(I_A + 1)\lambda} = \frac{m\rho - l}{(I_A + 1)m\rho}$$

is a linear function of l. The normal approximation to $f_Q(l)$ would have a maximum at $l = m\rho = \lambda/\mu$. The Poisson distribution, however, is slightly skewed and actually has a maximum at the nearest integer to $\lambda/\mu - 1/2$ as one can readily see from the relation

$$\frac{p_{j+1}}{p_j} = \frac{m\rho}{j+1}.$$

If $m\rho$ is an integer, then $p_{m\rho} = p_{m\rho - 1}$.

For $m\rho \gg 1$, we could approximate $f_Q(l)$ by

$$f_Q(l) \simeq C' \exp\left[\frac{-(l - m\rho + \frac{1}{2})^2}{m\rho(1 + I_A)} + \frac{2(l - m\rho + \frac{1}{2})^3}{3[m\rho(1 + I_A)]^2} + \ldots\right] \quad (8.7)$$

for $0 < l < m$, which gives the next correction to the normal approximation, a distribution which is slightly skewed and shifted.

We are mostly interested in the variance of the normal approximation which has the value $m\rho(1 + I_A)/2$. This is, of course, consistent with the variance for the M/M/n queue $I_A = 1$ for which the variance is $m\rho$. We have also evaluated Var $\{Q_s\}$ for a fairly general infinite-channel system in Section 6.4. In (6.28) we obtain for the exponential service distribution.

$$1 - C_S^* = \int_0^\infty e^{-2\mu x} dx \Big/ \int_0^\infty e^{-\mu x} dx = 1/2.$$

and

$$\frac{\text{Var}\{Q_s(t)\}}{E\{Q_s(t)\}} = 1 + (I_A - 1)\frac{1}{2} = \frac{1 + I_A}{2}$$

$$\text{Var}\{Q_s(t)\} = m\rho(1 + I_A)/2.$$

The variance obtained here is therefore consistent with that derived in Section 6.4 by quite different methods.

The unnormalized distribution (8.6) for exponentially distributed service times is obtained by taking the distribution for the infinite-channel system over $0 \leq l \leq m$ and joining it smoothly to the exponential distribution associated with a single server having arrival variance rate $I_A\lambda$ and departure variance rate μm. For $c < \infty$ the distribution is then truncated and normalized.

Although we recognized that there might be problems in trying to use a diffusion approximation for service distributions other than the exponential, it is tempting to ask if one couldn't simply join an approximate normal distribution for $0 \le l \le m$ to an exponential distribution for $m \le l$ under more general conditions. It is clear that the diffusion approximation would be valid for values of l sufficiently large that the departure process is essentially that associated with m independent busy servers. Thus we expect

$$f_Q(l) \simeq C'' \exp\left[\frac{2(1-\rho)l}{I_A \rho + I_D}\right]$$

for some constant C'' and sufficiently large l even for $I_D \neq 1$. Also, if there is to be a normal approximation for $l \le m$ and most customers entering the server came directly from the arrival process rather than the queue, the normal distribution should have the parameters obtained in Section 6.4 for the infinite-channel server.

The exponential service time distribution is unique in that the fluctuations in arrivals and departures interact in such a way as to give an effective combined variance coefficient of $I_A + 1$ or $I_A \rho + 1$ (which are essentially the same) for both $l > m$ and $l < m$. The result of this, in effect, is that the normal approximation for $l < m$ and the exponential distribution for $l > m$ join smoothly at $l \simeq m$. For nonexponential service time distributions, however, the interaction between the arrival and departure fluctuations are quite different for $l < m$ and $l > m$, and are quite complex for l close to m. Indeed, for $I_D \neq 1$, the correct distribution for $f_Q(l)$ is not simply a smooth joining of the obvious exponential and normal approximations.

As yet, no one has found any very satisfactory approximation to $f_Q(l)$ when $I_D \neq 1$ and ρ is in a range such that neither $P\{Q > m\}$ nor $P\{Q < m\}$ is small. If either of these is small, the queue behaves approximately like an effective single-channel server or an infinite-channel server, respectively. In general, the shape of $f_Q(l)$ will depend upon the $F_S(t)$.

8.2 Transient behavior, $\mu = \lambda$

Although the diffusion approximation gives some useful qualitative properties of equilibrium distributions, its accuracy is quite limited by the requirement that the (average) queue length be large compared with 1, which usually means that $1 - \rho \ll 1$. Since exact equilibrium distributions can be obtained for so many queueing systems, the diffusion approximation does not add very much to the class of (approximately) soluble equilibrium systems. The main advantage of diffusion approximations is that they also describe approximately the transient behavior of systems with constant arrival and departure rates, the behavior of some systems with time-dependent arrival and/or service rates, and some systems involving more than one dimension (tandem queues, several customer types, etc.). The diffusion approximations are most accurate when the queue stays away from boundaries for a time large compared with the typical interarrival or service time (for any reason). 'Exact solutions' of any of these time-dependent problems, in the few cases in which they have been obtained, are usually so complex as to be computationally useless.

We consider first the simple queue with constant arrival and service rate but starting from an arbitrary initial queue. If

$$\beta_x = \lambda, \quad \beta_y = \mu, \quad \beta_{xx} = I_A\lambda, \quad \beta_{yy} = I_D\mu \tag{8.8}$$

the diffusion equation becomes

$$\frac{\partial f_Q(l; t)}{\partial t} = \left[(\mu - \lambda)\frac{\partial}{\partial l} + \frac{1}{2}(I_A\lambda + I_D\mu)\frac{\partial^2}{\partial l^2} \right] f_Q(l; t) = 0. \tag{8.9}$$

Both $f_Q(l; t)$ and $F_Q(l; t)$ satisfy the same equation with different boundary conditions.

Solutions of (8.9) can be obtained in many different forms for any boundary conditions at $l = 0$ and $l = c$ but one can infer many properties of the solution directly from (8.9) without actually solving the equation.

First of all, if, for $c < \infty$, we let $l' = c - l$ then $f_Q(c - l'; t)$, $F_Q(c - l'; t)$ or $1 - F_Q(c - l'; t)$ each satisfy the equation

$$\frac{\partial f_Q(c - l'; t)}{\partial t} = \left[-(\mu - \lambda)\frac{\partial}{\partial l'} + \frac{1}{2}(I_A\lambda + I_D\mu)\frac{\partial^2}{\partial l'^2} \right] f_Q(c - l'; t) = 0, \tag{8.10}$$

which has the same form as (8.9) except for the sign of the $\partial/\partial l$ term. Furthermore, if no probability can stay on the boundary $1 - F_Q$ satisfies the boundary conditions

$$1 - F_Q(c - l'; t) = \begin{cases} 0 & \text{at } l' = 0 \\ 1 & \text{at } l' = c \end{cases}$$

the same boundary conditions as for $F_Q(l; t)$. This shows again the symmetry between customers and vacancies (which applies also to time-dependent behavior). Any solution $F_Q(l; t)$ of (8.9) for the distribution function of $Q(t)$ also gives a solution $1 - F_Q(c - l'; t)$ of (8.10) for the distribution function of vacancies $c - Q(t)$. But if we reinterpret vacancies as customers, the solution of (8.10) describes the evolution of another queueing system with arrival rate μ and service rate λ, and variance rates I_D and I_A, respectively. Thus, if $c < \infty$, there is for every solution with $\rho < 1$ a solution of a companion system with $\rho > 1$, i.e., $\rho \rightarrow 1/\rho$.

In the continuum approximation we can measure lengths in any units we wish. We can also measure time in any units. If we measure time in multiples of some arbitrary unit T_0 and queue lengths in multiples of some scale L_0, i.e., let

$$l^* = l/L_0 \quad \text{and} \quad t^* = t/T_0$$

and define

$$Q^*(t^*) = Q(t^* T_0)/L_0,$$

then $Q^*(t^*)$ would have a distribution function

$$F_{Q^*}(l^*; t^*) = P\{Q^*(t^*) \le l^*\} = P\{Q(t^* T_0) \le l^* L_0\}$$
$$= F_Q(l^* L_0; t^* T_0)$$

or

$$F_Q(l; t) = F_{Q^*}(l/L_0; t/T_0).$$

If F_Q satisfies (8.9), then F_{Q^*} will satisfy the equation

$$\frac{\partial F_{Q^*}(l^*; t^*)}{\partial t^*} = \left[\frac{T_0(\mu - \lambda)}{L_0} \frac{\partial}{\partial l^*} + \frac{1}{2} \frac{T_0(I_A \lambda + I_D \mu)}{L_0^2} \frac{\partial^2}{\partial l^{*2}} \right] F_{Q^*}(l^*; t^*)$$

$$(8.11)$$

in which the new coefficients represent the rate of growth of the queue and the variance rate in the new units. We have two free parameters T_0 and L_0 and two coefficients. We could choose units so as to give these coefficients (almost) any value we wish. The case $\mu = \lambda$, however, is special. The first derivative term vanishes no matter what the units are, but any change in units such that $L_0^2 = T_0$ leaves the equation unchanged.

The fact that, for $\lambda = \mu$, the differential equation is invariant to any changes in units such that $L_0^2 = T_0$ means that if $F_Q(l; t)$ is a solution of (8.9) with $\mu = \lambda$ so is $F_Q(l/L_0, t/L_0^2)$ for any L_0. If one can create some initial conditions and boundary conditions which are also invariant to changes in units and for which (8.9) has a unique solution, then this solution would have to satisfy

$$F_Q(l; t) = F_Q(l/L_0; t/L_0^2) \quad \text{for all } L_0,$$

in particular if we choose $L_0 = l$, then

$$F_Q(l; t) = F_Q(1, t/l^2) = h(l/t^{1/2}) \tag{8.12}$$

is some function only of $l/t^{1/2}$.

Actually there are such solutions. For example, the boundary conditions

$$F_Q(l; t) = \begin{cases} 1 & \text{for } l \to +\infty \\ 0 & \text{for } l = 0 \text{ (or } l = -\infty) \end{cases} \quad \text{for all } t$$

$$F_Q(l, 0) = \begin{cases} 1 & \text{for } l > 0 \\ 0 & \text{for } l < 0 \end{cases}$$

are independent of the units for l or t. These conditions correspond to an initial condition $Q(0) = 0$ and the boundary condition $0 \le Q(t) < +\infty$ (or $-\infty < Q(t) < +\infty$), i.e., $c = \infty$.

Indeed one can substitute (8.12) into (8.9) and verify that (8.12) is a solution of (8.9) for $\mu = \lambda$ provided $h(z)$ satisfies the ordinary differential equation

$$-z \frac{dh}{dz} = (I_A + I_D)\mu \frac{d^2h}{dz^2}.$$

The solution of this first order equation for dh/dz gives

$$\frac{dh}{dz} = C \exp\left\{ -\frac{z^2}{2(I_A + I_D)\mu} \right\}.$$

The particular solutions of interest are

$$h(z) = 2\Phi\left(\frac{z}{[(I_A + I_D)\mu]^{1/2}} \right) - 1 \quad \text{or} \quad \Phi\left(\frac{z}{[(I_A + I_D)\mu]^{1/2}} \right).$$

giving

$$F_Q(x; t) = 2\Phi\left[\frac{l}{[(I_A + I_D)\mu t]^{1/2}} \right] - 1$$

$$= \Phi\left[\frac{l}{[(I_A + I_D)\mu t]^{1/2}} \right] - \Phi\left[\frac{-l}{[(I_A + I_D)\mu t]^{1/2}} \right]$$

$$\text{or } \Phi\left[\frac{l}{[(I_A + I_D)\mu t]^{1/2}} \right].$$

for the boundary condition $0 \le Q(t) < \infty$ or $-\infty \le Q(t) < \infty$, respectively.

The solution for no boundary condition at $l = 0$, $-\infty < Q(t) < +\infty$, comes as no surprise since this normal distribution is a solution from which the diffusion equation was derived. We have merely shown how this could have been inferred from the dimensional symmetry of (8.9). For $\mu = \lambda$, however, (8.9) is also invariant to changing l to $-l$. If $G(l; t)$ is any solution of (8.9) so is $G(-l; t)$ or any linear combination of $G(l; t)$ and $G(-l; t)$. In particular

$$F_Q(l; t) = G(l; t) - G(-l; t)$$

and

$$f_Q(l; t) = G(l; t) + G(-l; t)$$

are solutions. These antisymmetric and symmetric solutions automatically satisfy the conditions

$$F_Q(l; t) \to 0 \quad \text{for } l \to 0 \quad \text{for all } t$$

and

$$\frac{\partial f_Q(l; t)}{\partial l} \to 0 \quad \text{for } l \to 0.$$

The above solution for $0 \le Q(t) < \infty$ could indeed have been obtained by choosing $G(l; t)$ as the known solution for $-\infty < Q(t) < +\infty$. The corresponding probability density $f_Q(l; t)$ for $0 \le Q(t) \le \infty$ could similarly be obtained by choosing $G(l; t)$ as the normal probability density for $-\infty < Q(t) < +\infty$.

Equation (8.9) has another obvious symmetry. It is invariant to translations in either l or t. If $F_Q(l; t)$ is a solution of (8.9) so is $F_Q(l - l_0; t - t_0)$ for any l_0 or t_0. The solution of (8.9) with $\mu = \lambda$ satisfying the initial condition $Q(t_0) = l_0$ and the boundary condition $0 \le Q(t) < \infty$ is therefore

$$F_Q(l; t) = \Phi\left[\frac{l - l_0}{[(I_A + I_D)\mu(t - t_0)]^{1/2}}\right]$$
$$- \Phi\left[\frac{-l - l_0}{[(I_A + I_D)\mu(t - t_0)]^{1/2}}\right]$$
$$= \Phi\left[\frac{l - l_0}{[(I_A + I_D)\mu(t - t_0)]^{1/2}}\right]$$
$$+ \Phi\left[\frac{l + l_0}{[(I_A + I_D)\mu(t - t_0)]^{1/2}}\right] - 1,$$

having a probability density

$$f_Q(l; t) = \frac{1}{[2\pi(I_A + I_D)\mu(t - t_0)]^{1/2}} \left\{ \exp\left[\frac{-(l - l_0)^2}{2(I_A + I_D)\mu(t - t_0)} \right] \right.$$

$$\left. + \exp\left[\frac{-(l + l_0)^2}{2(I_A + I_D)\mu(t - t_0)} \right] \right\}. \tag{8.13}$$

This solution is often described as an 'image' solution. If the first term of (8.13) is interpreted as the mass distribution resulting from a unit mass being placed at l_0 at time t_0, the second term can be interpreted as the mass distribution resulting from placing an image source at $-l_0$ (behind the mirror). Of the total mass of two units on $-\infty \leq l < +\infty$, only one unit stays in the range $0 \leq l < \infty$, for all t.

Although this solution for $\mu = \lambda$ and $c = \infty$ is rather specialized, it shows some important properties that are typical of the more complex solutions with $\mu \neq \lambda$ and/or $c < \infty$. Since for sufficiently small $t - t_0$ the behavior of the queue is dominated by the fluctuations rather than the mean drift, the effect of the term of (8.9) proportional to $\mu - \lambda$ will be negligible. If, in particular, we start with no queue, $Q(t_0) = 0$.

$$E\{Q(t) | Q(t_0) = 0\} = \int_0^\infty l f_Q(l; t) dl = [2(I_A + I_D)\mu(t - t_0)/\pi]^{1/2},$$

the queue grows proportional to $(t - t_0)^{1/2}$.

The solution of (8.9) with an absorbing boundary at $l = 0$ is equally simple for $\mu = \lambda$; we need only impose the boundary condition that $f_Q(l; t) = 0$ at $l = 0$, instead of $F_Q(l; t) = 0$. If, instead of taking the symmetric function (8.13), we took the corresponding antisymmetric function, $f_Q(l; t)$ would represent the probability density for $Q(t)$ given that $Q(t_0) = l_0$ and that $Q(\tau) > 0$ for $t_0 < \tau < t$. By integrating this over all $l > 0$, we conclude that

$$P\{Q(\tau) > 0 \text{ for } t_0 < \tau < t\} = 2\Phi\left[\frac{l_0}{[(I_A + I_D)\mu(t - t_0)]^{1/2}} \right] - 1. \tag{8.14}$$

If Y_0 denotes the first time that $Q(t) = 0$, (8.14) also represents $P\{Y_0 > t | Q(t_0) = l_0\}$. This is a function only of $(t - t_0)/l_0^2$, thus a change in l_0, in effect, merely changes the units for $t - t_0$.

If $(t - t_0)(I_A + I_D)\mu/l_0^2 \gg 1$

$$P\{Y_0 > t \mid Q(t_0) = l_0\} \simeq \frac{l_0}{[\pi(I_A + I_D)\mu(t - t_0)/2]^{1/2}}.$$

Although Y_0 has a proper distribution function, $P\{Y_0 > t\} \to 0$ for $t \to \infty$, the tail decreases very slowly, like $(t - t_0)^{-1/2}$, so as to give $E\{Y_0\} = \infty$.

For $c < \infty$ and $\mu = \lambda$, the solution of (8.9) cannot generally be obtained in closed form but infinite series solutions can be obtained in either of two forms, one of which converges rapidly for small $t - t_0$ and the other for large $t - t_0$.

For $l_0/c \ll 1$, the mass distribution will 'feel the boundary' at $x = 0$ before it has had time to reach the boundary at c; consequently, (8.13) will describe $f_Q(l; t)$ at least until there is a significant probability that $Q(t)$ can reach c. On the other hand, if l_0 is close to c, the mass distribution feels the boundary at c before it can reach $l = 0$. By virtue of the symmetry between customers and vacancies, $f_Q(c - l; t)$ satisfies the same equation as $f_Q(l; t)$, therefore (8.13) with l and l_0 replaced by $c - l$ and $c - l_0$ respectively would describe the evolution of $f_Q(l; t)$ at least until $Q(t)$ can reach 0. The latter solution could also be obtained by placing an image mass above c at $c + (c - l_0)$.

The general solution with reflecting boundaries at both $l = 0$ and c can be obtained by reflecting the source at l_0 and its images back and forth over the two boundaries or equivalently by placing a unit source at each of the points

$$\pm l_0 + 2kc \quad \text{for} \quad k = 0, \pm 1, \pm 2, \ldots$$

(periodic with period $2c$). The complete solution can thus be written as an infinite series.

$$\begin{aligned}
f_Q(l; t) = \sum_{k = -\infty}^{+\infty} &\frac{1}{[2\pi(I_A + I_D)\mu(t - t_0)]^{1/2}} \\
&\times \left\{ \exp\left[\frac{-(l - l_0 - 2kc)^2}{2(I_A + I_D)\mu(t - t_0)} \right] \right. \\
&\left. + \exp\left[\frac{-(l + l_0 - 2kc)^2}{2(I_A + I_D)\mu(t - t_0)} \right] \right\}
\end{aligned} \qquad (8.15)$$

each term of which (except the primary source at l_0) is the mass distribution generated by one of the image sources.

This series certainly converges for any finite value of $(t - t_0)$ because it takes a longer and longer time for mass from image sources further away to reach the interval $0 < l < c$. This form of the solution, however, is convenient only if at most a few terms of (8.15) contribute significantly to the sum. If, for example,

$$\frac{(2c)^2}{2(I_A + I_D)(t - t_0)} \gtrsim 1$$

i.e.,

$$(t - t_0) \lesssim \frac{2c^2}{(I_A + I_D)\mu} \tag{8.16}$$

the terms with $|k| > 1$ will contribute very little to (8.15).

For larger values of $(t - t_0)$ it is most convenient to solve (8.9) by Fourier series. For $\lambda = \mu$, the solution of (8.9) with $\partial f_Q/\partial l = 0$ at $l = 0$ and $l = c$, and $Q(t_0) = l_0$ can be written in the form

$$f_Q(l; t) = \frac{1}{c} + \frac{2}{c} \sum_{j=1}^{\infty} \cos\left[\frac{j\pi l}{c}\right] \cos\left[\frac{j\pi l_0}{c}\right] \exp\left[-\lambda_j(t - t_0)\right] \tag{8.17}$$

with

$$\lambda_j = \frac{(I_A + I_D)\mu j^2 \pi^2}{2c^2}.$$

The first term of (8.17) is, of course, the equilibrium distribution for $\lambda = \mu$ and succeeding terms describe the transient behavior. Since $\lambda_j/\lambda_1 = j^2$ increases rapidly with j, this series converges very rapidly for

$$t - t_0 \gtrsim \frac{1}{\lambda_1} = \frac{2c^2}{(I_A + I_D)\mu\pi^2}. \tag{8.18}$$

For $t - t_0 > 1/\lambda_1$, the transient terms decay exponentially fast in time with the 'time constant' $1/\lambda_1$. For $l_0 = c/2$, however, the term for $j = 1$ vanishes. The leading term for $j = 2$ decays even faster with a time constant of $1/4\lambda_1$.

Fortunately the two ranges (8.16) and (8.18) overlap very generously. By the time $Q(t)$ has a significant probability of moving even a distance of c so as to feel both boundaries, the series (8.17) is already converging very rapidly. If l_0 is close to $c/2$, $Q(t)$ will feel both boundaries by the time it can travel only $c/2$.

Analogous series solutions for $f_Q(l; t)$ can be obtained if $l = 0$ and/or $l = c$ is an absorbing boundary.

8.3   Transient behavior, $\mu \neq \lambda$

In the special case $\mu = \lambda$ discussed in the last section, (8.9) itself
defined no natural units of length or time, but it did define a unit of
(length)^2/(time), namely $(I_A + I_D)\mu$. The only lengths in the problem
were imposed by the initial condition l_0 and the boundary c (if finite).
The consequence of this was that, for $c = \infty$, the solution (8.13),
except for a scaling factor, was a function only of the dimensionless
combinations

$$l/l_0 \quad \text{and} \quad (t - t_0)\left[(I_A + I_D)\mu/l_0^2\right].$$

For $c < \infty$, the solution (8.15) or (8.17) was a function only of these
variables plus l_0/c or, equivalently of

$$l_0/c, \quad l/c, \quad \text{and} \quad (t - t_0)\left[(I_A + I_D)\mu/c^2\right].$$

Two symmetries which were exploited to solve (8.9) for $\mu = \lambda$ no
longer apply for $\mu \neq \lambda$; the invariance to a change in units with L_0^2
$= T_0$ in (8.11), and to reflections $l \to -l$. We can use the freedom of L_0
and T_0, however, to create a dimensionless form of (8.11). If, as in
Section 5.2, we choose

$$T_0 = \frac{I_A\rho + I_D}{|1 - \rho|^2}\frac{1}{\mu}, \quad L_0 = \frac{I_A\rho + I_D}{|1 - \rho|}$$

then (8.11) simplies to

$$\frac{\partial F_{Q^*}(l^*, t^*)}{\partial t^*} = \left[\pm\frac{\partial}{\partial l^*} + \frac{1}{2}\frac{\partial^2}{\partial l^{*2}}\right]F_{Q^*}(l^*; t^*) \qquad (8.19)$$

in which the upper sign applies for $\rho < 1$, the lower sign for $\rho > 1$.
   For $c < \infty$ and $c^* \equiv c/L_0$, the transformation $l^* \to c^* - l^*$ will
reverse the sign in (8.19) changing customers into vacancies and
$\rho \to 1/\rho$ (or vice versa). Thus any solution with $\rho > 1$ can be mapped
into a solution for $\rho < 1$.
   Even if we could not solve (8.19) analytically, we have already
greatly simplified the problem. We have shown that the time-
dependent solutions $F_Q(l; t)$ of (8.9) for various values of λ, μ, I_A and
I_D differ only by a change in units. If we can solve any time-dependent
queueing problem of the type (8.19) with any values of these
parameters by simulation, numerical integration, or any other
scheme, we will also have solved a whole family of queueing problems
for any other values of λ, μ, ... The solution of (8.19) will, however,
still depend upon the initial queue l_0^*, and c^* (if $c^* < \infty$).

Actually the solution of (8.19) can be obtained in several forms by a slight variation on the methods used in the last section for $\mu = \lambda$. In the absence of any boundaries the solution of (8.19) is (for $t_0 = 0$)

$$F_{Q^*}(l^*; t^*) = \Phi\left(\frac{l^* - l_0^* \pm t^*}{\sqrt{t^*}}\right)$$

or

$$f_{Q^*}(l^*; t^*) = \frac{1}{(2\pi t^*)^{1/2}} \exp\left(-\frac{(l^* - l_0^* \pm t^*)^2}{2t^*}\right)$$

$$= \frac{1}{(2\pi t^*)^{1/2}} \exp\left(-\frac{(l^* - l_0^*)^2}{2t}\right)$$

$$\times \exp\left[\mp (l^* - l_0^*)\right] \exp\left(-\frac{t^*}{2}\right)$$

which describes a diffusion in a coordinate system moving with velocity ∓ 1 (in the new units) as illustrated in Fig. 4.5.

The last factored form of f_{Q^*} is difficult to interpret 'physically' but it does suggest a useful mathematical trick. The first factor is a solution of the diffusion equation with no drift and the other factors are simple exponentials in l^* and t^*. If we let

$$F_{Q^*}(l^*; t^*) = e^{\mp l^*} G(l^*; t^*)$$

then G would satisfy the equation

$$\frac{\partial G(l^*; t^*)}{\partial t^*} = \frac{1}{2}\left(-1 + \frac{\partial^2}{\partial l^{*2}}\right) G(l^*; t^*). \tag{8.20}$$

This transformation has eliminated the first derivative term, $\pm \partial/\partial l^*$, from (8.19) but replaced it by a constant term $-1/2$ in (8.20). We could also have eliminated this constant term by factoring out an $e^{-t/2}$ from F_{Q^*}, but this is of no additional advantage.

What we have gained by this is an equation (8.20) which is invariant to changing l^* to $-l^*$ and to which we can again apply image methods. If $G_0(l^*; t)$ is any solution of (8.20), so is $G_0(l^*; t^*) \pm G_0(-l^*; t^*)$. It follows that if $F_{Q^*}(l^*; t^*)$ is any solution of (8.19), so is

$$F_{Q^*}(l^*; t^*) - e^{\mp 2l^*} F_{Q^*}(-l^*; t^*)$$

and, in particular,

$$F_{Q^*}(l^*; t^*) = \Phi\left(\frac{l^* - l_0^* \pm t^*}{\sqrt{t^*}}\right) - e^{\mp 2l^*} \Phi\left(\frac{-l^* - l_0^* \pm t^*}{\sqrt{t^*}}\right) \tag{8.21}$$

is the solution of (8.19) which satisfies the boundary conditions

$$F_{Q^*}(0; t^*) = 0 \quad \text{and} \quad F_{Q^*}(l^*; t^*) \to 1 \quad \text{for } l^* \to +\infty$$

with $Q^*(0) = l_0^*$, i.e., the solution for $c^* = \infty$.

This rather simple formula actually describes quite a few effects. The first term of (8.21), of course, describes the diffusion with no boundary. For $\rho < 1$ (upper sign) the mass center of this term is at $l_0^* - t^*$, moving with a velocity -1. The center hits the boundary at $t^* = l_0^*$ and by time t^* with $t^* - l_0^* \gtrsim t^{*1/2}$ most of this mass will have crossed the boundary causing the first term of (8.21) to approach 1 for all $l^* > 0$.

The Φ-function in the second term of (8.21), or actually $1 - \Phi$, represents the distribution function of a hypothetical mass starting at an image point $-l_0^*$ and moving in the opposite direction with a velocity $+1$ for $\rho < 1$. It passes through the boundary from below and is just the reflection of the behavior of the first term. For $t^* - l_0^* \gtrsim t^{*1/2}$ most of this mass will have traveled into the range $l^* > 0$ and for $t^* \to \infty$ the second Φ-function also goes to 1 for any $l^* < \infty$ giving the equilibrium distribution

$$1 - F_{Q^*}(l^*; t^*) \to e^{-2l^*} \quad \text{for } t^* \to \infty. \tag{8.22}$$

The approach to the equilibrium distribution, however, depends upon l_0^*. For $l_0^* \ll 1$ most of the mass feels the boundary very quickly. The first term of (8.21) goes to 1 rapidly with t^* (for t^* on a scale of order 1) and the Φ-function in the second term also goes to 1 quite rapidly for $l^* = 0$. It takes additional time, however, for the second mass distribution traveling with a mean velocity $+1$ to reach some $l^* > 0$. The consequence of this is that, for any finite t, the tail of the limit distribution (8.22) for $l^* \gtrsim t^*$ is slower to develop than the part for $l^* \lesssim t^*$. Fig. 8.1 shows the distribution $1 - F_{Q^*}(l^*; t^*)$ for $l_0^* = 0$ and several values of t^*.

For $l_0^* \gg 1$, it takes a time $t^* = l_0^* \gg 1$ for the center of the mass distribution to reach the boundary, by which time the mass distribution has also spread on a scale of order $t^{*1/2} \simeq l_0^{*1/2} \gg 1$. Any mass which can reach the boundary by a time t^* will be redistributed into the equilibrium distribution within a time of order 1 later, a time which is short compared with the time of order $l_0^{*1/2}$ that it takes the whole mass distribution to cross the boundary. Thus the mass that has gone into the equilibrium distribution by time t^* must be approximately that which has crossed the boundary. Analytically, what this means is that over the (small) range of l^* of order 1 where

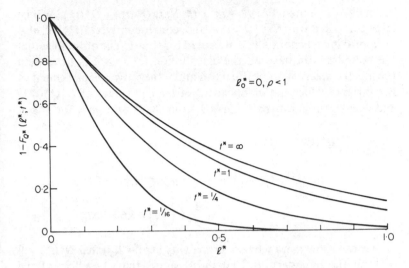

Figure 8.1 *Transient queue distribution for $Q(0)$ and $\rho < 1$*

$\exp(-2l^*)$ is significant, the second Φ-function in (8.21) is nearly independent of l^* (for $l^* > 0$) and consequently

$$F_{Q^*}(l^*; t^*) \simeq \Phi\left(\frac{l^* - l_0^* + t^*}{\sqrt{t^*}}\right) - e^{-2l^*}\Phi\left(\frac{-l_0^* + t^*}{\sqrt{l_0^*}}\right) \quad (8.23)$$

for $l_0^* \gg 1$ and $\rho < 1$.

For $\rho > 1$ (lower sign) the mass center of the first term of (8.21) is moving with a velocity $+1$, away from the boundary, but some of the mass could reach the boundary before it finally escapes. The mass center of the second Φ-function starting at $-l_0^*$ is moving with a velocity -1 also away from the boundary. The second term, however, is rather awkward to evaluate numerically because it is the product of a rapidly increasing function of l^*, $\exp(2l^*)$, and an even more rapidly decreasing function of l^*. To evaluate (8.21) for $l^* \gg 1$ one could use asymptotic approximations for the Φ-function in the second term, but there is a simpler more intuitive way of approximating it.

Even if we start with $Q^*(0) = 0$, there will be a time t_0^* (of order 1) for which $Q^*(t_0^*)$ is sufficiently large that it is virtually impossible for $Q^*(t^*)$ to reach the boundary for any $t^* > t_0^*$. The queue at time t^*, however, can be expressed as the sum of $Q^*(t_0^*)$ and the change in the queue, $Q^*(t^*) - Q^*(t_0^*)$, during the time $t^* - t_0^*$. These two random variables are nearly independent and the latter is approximately

normally distributed. For $t^* - t_0^* \gg t_0^*$, Var $\{Q^*(t^*) - Q^*(t_0^*)\}$ will be
large compared with Var $\{Q^*(t_0^*)\}$ and, consequently, $Q^*(t^*)$ will also
be approximately normally distributed for $t^* \gg 1$. The only effect that
the boundary can have on the distribution for $t^* \gg 1$ is to cause a
(positive) displacement of the mean and a (negative) displacement of
the variance which are independent of t^*. The values of $E\{Q^*(t^*)\}$
and Var $\{Q^*(t^*)\}$ can be evaluated from (8.21) and give, for $t^* \gg 1$,

$$E\{Q^*(t^*)\} - (l_0^* + t^*) \simeq \frac{1}{2}\exp(-2l_0^*) \tag{8.24}$$

$$\mathrm{Var}\{Q^*(t^*)\} - t^* \simeq -\frac{1}{2}(1 + 2l_0^*)\exp(-2l_0^*)$$
$$-\frac{1}{4}(1 + 4l_0^*)\exp(-4l_0^*).$$

One can show that there is a probability $\exp(-2l_0^*)$ that $Q^*(t^*)$ will
ever hit the boundary. Indeed (8.24) shows that the effect of the
boundary on $E\{Q^*(t^*)\}$ is also proportional to $\exp(-2l_0^*)$. The
effect on Var $Q^*(t^*)$ is more complicated but is also dominated by the
same factor for $l_0^* \gg 1$. For $l_0^* = 0$

$$E\{Q^*(t^*)\} \simeq t^* + \frac{1}{2} = 5/4 + (t^* - 3/4)$$
$$\mathrm{Var}\{Q^*(t^*)\} \simeq (t^* - 3/4) \tag{8.25}$$

so $Q^*(t^*)$ eventually behaves as if it started at $Q^*(t_0^*) = 5/4$ at
$t_0^* = 3/4$ instead of $Q^*(0) = 0$ at time 0.

Fig. 8.2 shows the distribution $1 - F_{Q^*}(l^*; t^*)$ for several values of
t^*. For $t^* = 1/16$ the curve in Fig. 8.2 for $\rho > 1$ is not very different
from that of Fig. 8.1 for $\rho < 1$ (although these are drawn on different
scales). For short times, the behavior of the queue is dominated by the
fluctuations and their interaction with the boundary; the mean drift of
$\mp 1/16$ in a time $1/16$ is a secondary effect.

The two broken line curves represent the queue distribution that
would exist in the absence of a boundary for $t^* = 1$, $1 - \Phi(l^* - 1)$, and
the normal approximation (8.25), $1 - \Phi(2l^* - 3)$, with $t^* = 1$.
Obviously $t^* = 1$ is not long enough for the latter approximation to
be very accurate, but this at least illustrates the trend.

To obtain the queue distribution for $\mu = \lambda$ and $c < \infty$ we made
successive reflections of the probability density over the boundaries at
0 and c because the formulas for the density seemed more appealing.
For $\mu \neq \lambda$, however, the derivative of (8.23) with respect to l^*
generates three terms and, furthermore, the boundary conditions for

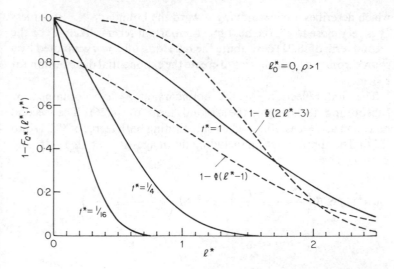

Figure 8.2   *Transient queue distribution for $Q(0) = 0$ and $\rho > 1$*

the density at $l^* = 0$ and c^* are not as simple as those for the distribution function. It suffices to consider $\rho < 1$ since the corresponding solutions for $\rho > 1$ can be obtained by interchanging customers and vacancies.

We could derive an infinite series solution generalizing (8.15) but it suffices to consider only the first few reflections because, as for $\mu = \lambda$, a Fourier series expansion will converge very rapidly by the time the distribution has had time to reflect off both boundaries.

If $1 - F_{Q^*}(l^*; t)$ is any complementary distribution function, also a solution of (8.19), then so is

$$[1 - F_{Q^*}(l^*; t^*)] - e^{2(c^* - l^*)}[1 - F_{Q^*}(2c^* - l^*; t^*)].$$

(This is essentially the distribution function of the vacancies.) Furthermore, this vanishes at $l^* = c^*$. Therefore,

$$F_{Q^*}(l^*; t^*) + e^{2(c^* - l^*)}[1 - F_{Q^*}(2c^* - l^*; t^*)] \tag{8.26}$$

has the value 1 at $l^* = c$ for any solution $F_{Q^*}(2c^* - l^*; t^*)$ of (8.19).

For sufficiently small t^*, before $Q^*(t^*)$ can reach either boundary (for $0 < l_0^* < c^*$), the distribution starting from $Q^*(0) = l_0^*$ is primarily just the first term of (8.21)

$$F_{Q^*}(l^*; t) \simeq \Phi\left(\frac{l^* - l_0^* + t^*}{\sqrt{t^*}}\right) \tag{8.27}$$

which describes a mass moving toward the boundary $l^* = 0$. Unless l_0^* is very close to c^*, the next most important term is likely to be the second term of (8.21) describing the reflection off $l = 0$ generated by a source from $-l_0^*$ which tries to create the exponential distribution for $c = \infty$.

The first reflection off the boundary at $l = c$ is obtained by substituting (8.27) into the second term of (8.26). The second reflection at $l = c$ is obtained by substituting both terms of (8.21) into (8.26). The approximation including the images at $-l_0^*$, $2c^* - l_0^*$, and $2c^* + l_0^*$ gives

$$F_{Q^*}(l^*; t) \simeq \Phi\left(\frac{l^* - l_0^* + t^*}{\sqrt{t^*}}\right) - e^{-2l^*}\Phi\left(\frac{-l^* - l_0^* + t^*}{\sqrt{t^*}}\right)$$
$$+ e^{-2l^* + 2c^*}\Phi\left(\frac{l^* - 2c^* + l_0^* - t^*}{\sqrt{t}}\right)$$
$$+ e^{-2c^*}\Phi\left(\frac{-2c^* + l^* - l_0^* + t^*}{\sqrt{t^*}}\right). \tag{8.28}$$

This satisfies exactly the boundary condition at $l^* = c^*$ but the last two terms do not quite satisfy the boundary conditions at $l = 0$. The next approximation with images at $-2c^* + l_0^*$ and $-2c^* - l_0^*$ would satisfy the boundary conditions at $l^* = 0$ but not at $l^* = c^*$, etc.

For sufficiently large t^*, it may be advantageous to solve (8.19) by Fourier series. The equilibrium distribution (8.2) is

$$F_{Q^*}(l^*) = \frac{1 - e^{-2l^*}}{1 - e^{-2c^*}}$$

and $F_{Q^*}(l^*; t^*) - F_{Q^*}(l^*)$ also satisfies (8.19) but with the boundary conditions that it vanishes at both $l^* = 0$ and $l^* = c^*$. Therefore

$$G(l^*; t) = e^{l^*}[F_{Q^*}(l^*; t^*) - F_{Q^*}(l^*)]$$

would be a solution of (8.20) which vanishes at $l^* = 0$ and c^* for all t^*. The product solutions of (8.20), which vanish at $l^* = 0$ and c^* are

$$\exp(-\lambda_j^* t^*)\sin\left(\frac{j\pi l^*}{c^*}\right), \quad j = 1, 2, \ldots$$

with

$$\lambda_j^* = \tfrac{1}{2} + \frac{j^2\pi^2}{c^{*2}}. \tag{8.29}$$

Therefore $F_{Q^*}(l^*; t^*)$ can be written in the form

$$F_{Q^*}(l^*; t) = \frac{1 - e^{-2l^*}}{1 - e^{-2c^*}} + \sum_{j=1}^{\infty} a_j(l_0^*) \exp(-\lambda_j^* t^*) e^{-l^*} \sin\left(\frac{j\pi l^*}{c^*}\right)$$

(8.30)

in which the $a_j(l_0^*)$ are determined so as to satisfy the initial conditions

$$F_{Q^*}(l^*; 0) = \begin{cases} 0, & 0 < l^* < l_0^* \\ 1, & l_0^* < l^* < c^* \end{cases}.$$

The $a_j(l_0^*)$ can be evaluated explicitly but they are uninteresting.

The solution (8.30) is exact for all possible choices of c^* and l_0^*. The series always converges but it converges rapidly only under just those conditions when the image method becomes awkward. There are actually four parameters in the original problem, however, c, l_0 and the two scale factors L_0 and T_0. If we convert the solution back to the original units, the time-dependent part of (8.30) will decay like

$$\exp(-\lambda_j^* t^*) = \exp(-\lambda_j t)$$

and

$$\lambda_j = \frac{1}{2T_0} + \frac{(I_A \rho + I_D) \mu j^2 \pi^2}{2c^2}$$

$$= \frac{\mu |1 - \rho|^2}{2(I_A \rho + I_D)} + \frac{(I_A \rho + I_D) \mu j^2 \pi^2}{2c^2}.$$

(8.31)

For $\rho = 1$, $T_0 = \infty$ and these λ_j are the same as occurred in the Fourier series expansion (8.17) for $\mu = \lambda$. The solution (8.17), however, will be approximately correct even for $\rho \neq 1$ provided that the first term of (8.31) is small compared with the second term (for $j = 1$). From (8.29) we see that this is true provided that $c^*/\pi \ll 1$, $c \ll \pi L_0$, i.e., the queue behavior will not recognize that $\rho \neq 1$ if the scale of the equilibrium queue length is large compared with the storage space.

If c^* is comparable with 1, we see from (8.31) that the transients decay somewhat faster for $\rho \neq 1$ than for $\rho = 1$ (for fixed values of the other parameters). The series (8.30) converges rapidly for $t \gtrsim 1/\lambda_1$ but, as for $\lambda = \mu$, the image solution (8.28) converges rapidly for $t \lesssim 1/\lambda_1$ (with a generous range where they both converge rapidly). For $t \gtrsim 1/\lambda_1$ the deviation from the equilibrium distribution will be dominated by the slowest decaying transient, the term for $j = 1$ (provided $a_1(l_0^*) \neq 0$). Whereas the equilibrium distribution function has a (truncated) tail that decreases with l^* like $\exp(-2l^*)$, the transient terms are proportional to $\exp(-l^*)$. Thus, for $\rho < 1$, the tail of the distribution (l^* near c^*) will appear to approach its limiting shape somewhat slower than for small l^*, in the sense that $\exp(-l^*)$ decays with l^* much slower than $\exp(-2l^*)$.

For $c^* \gg 1$, the transient terms of (8.30) all decay faster than $\exp(-t^*/2)$ but at nearly the same rate for all j with $j \ll c^*/\pi$. Whether the series (8.30) converges rapidly or not depends upon the coefficients $a_j(l_0^*)$; the factors $\exp(-\lambda_j^* t^*)$ will guarantee eventual convergence but decay of this factor will not become effective until j is comparable with c^*/π. On the other hand, the image solution (8.28) converges rapidly for all t^* and demonstrates when or why (8.30) might converge slowly.

For $c^* = \infty$, the image solution (8.23) is exact and leads to the equilibrium distribution

$$F_{Q^*}(l^*; t^*) \to 1 - e^{-2l^*} \quad \text{for} \quad t^* \to \infty.$$

The equilibrium distribution for $c^* < \infty$, however, differs from that for $c^* = \infty$ by a truncation and renormalization, but it can be expanded in powers of e^{-2c^*}

$$F_{Q^*}(l^*) = \frac{1 - e^{-2l^*}}{1 - e^{-2c^*}} = (1 - e^{-2l^*})(1 + e^{-2c^*} + e^{-4c^*} + \ldots)$$

$$(8.32)$$

which obviously will converge very rapidly even for $c^* \gtrsim 1$. Whereas the first term of (8.32) obtains from (8.23) as a result of reflection over $l^* = 0$ for $t^* \to \infty$, succeeding terms of (8.32) result from successive reflection of these terms over $l^* = c$ and $l^* = 0$. The series (8.28) converges rapidly for all t^* because the probability mass seldom reaches the boundary at $l^* = c^*$, expect possibly for small t^* if the queue was initially close to c^*.

The series (8.30), in fact, converges very slowly for $c^* \gg 1$ for all values of l_0^* because the form of the l^* dependence for the individual terms is not representative of the transient behavior. For $c^* \to \infty$, the series would actually be approximated by a Fourier integral rather than a series.

Problems

8.1 Approximate $f_Q(l; t)$ by the terms of (8.15) for $k = 0, \pm 1$ and draw a graph of this approximation as a function of l/c, $0 < l < c$ for

$$(t - t_0)(I_A + I_D)\mu/c^2 = 1/2, \quad l_0 = 0.$$

Compare this with a graph of (8.17) approximated by the term for only $j = 1$.

8.2 Verify by integration of (8.5) that

$$\int_0^c f_Q(l; t)\,dl = 1 \quad \text{for all } t.$$

8.3 If (8.21) is interpreted as the conditional distribution of $Q^*(t)$ given $Q^*(0) = l_0^*$, $F_{Q^*}(l^*; t^* | Q^*(0) = l_0^*)$, then for any distribution $F_0(l_0^*)$ of $Q^*(0)$, the distribution of $Q^*(t)$ can be written as

$$F_{Q^*}(l^*; t^*) = \int_0^\infty F_{Q^*}(l^*; t^* | l_0^*)\,dF_0(l_0^*).$$

If $F_0(l_0^*)$ is the equilibrium distribution

$$F_0(l_0^*) = 1 - e^{-2l_0^*}$$

(for $\rho < 1$), we expect that the distribution of $Q^*(t)$ will be independent of t^*, i.e.,

$$F_{Q^*}(l^*; t^*) = 1 - e^{-2l^*}.$$

Verify that (8.21) has this property.

CHAPTER 9

Time-dependent queues

9.1 Introduction

We are now prepared to analyze in more detail the time-dependent stochastic behavior of certain elementary queueing systems previously described in Chapter 2 via deterministic approximations, particularly the properties of a queue with one customer type and a single server during a rush hour (time-dependent arrival rate) and of a queue with pulsed service (or other time-dependent service rate).

We consider first the typical rush hour in which $\lambda(t)$ starts at some low value, increases to a maximum usually exceeding the service rate μ, and then decreases until the queue vanishes. The deterministic version of this was treated in Fig. 2.3, and some properties of typical realizations were discussed in Chapter 4.

We already know how to obtain crude approximations for at least some parts of the queue evolution. If $\lambda(t)$ increases sufficiently slowly when $\lambda(t)$ is less than μ that (5.7) is true, then in the deterministic approximation the average queue vanishes, but in the next approximation we could postulate that the queue distribution manages at all such times to stay close to the equilibrium distribution associated with the value of $\lambda(t)$ at the time t. If, for example, the queue distribution satisfies the diffusion equation (7.19) or (7.22) with

$$\beta_x = \lambda(t), \quad \beta_y = \mu, \quad \beta_{xx} = I_A \lambda(t), \quad \beta_{yy} = I_D \mu$$

then an approximate time-dependent solution of the equation would be

$$1 - F_Q(l; t) \simeq \exp(-2l/L_0(t)) \qquad (9.1)$$

with

$$L_0(t) = \frac{I_A \rho(t) + I_D}{1 - \rho(t)}, \quad \rho(t) = \lambda(t)/\mu.$$

As $\rho(t)$ approaches 1 and (9.1) fails, the typical queue cannot grow fast enough to keep up with the rapidly growing equilibrium queue length

$L_0(t)$. For $\rho < 1$, a positive queue will on the average decrease at a (slow) rate $\mu - \lambda(t)$ but fluctuations and reflections off the boundary $l = 0$ cause the average queue to increase. Even for $\rho = 1$, however, we know from Section 8.2 that the average queue will grow only in proportion to $\tau^{1/2}$ in a time τ.

We could follow the qualitative behavior of the queue by approximating $\lambda(t)$ by a piecewise constant function as it grows and passes through the critical range $\lambda(t) \simeq \mu$. Starting at some time τ_0 when the queue distribution is still close to the equilibrium distribution for $\rho(\tau_0)$ we could suddenly change $\rho(t)$ to some value $\rho(\tau_1), \tau_1 > \tau_0$ and evaluate the time-dependent (transient) queue behavior from this initial distribution until a time $\tau_2 > \tau_1$ using the transient solutions of Section 8.3. With this distribution at time τ_2 as a new initial distribution, we could increase $\rho(t)$ to some value $\rho(\tau_3)$ and follow the transient solution again until some time τ_4, etc. The time intervals $\tau_j - \tau_{j-1}$ need not be particularly short because, for $\rho(t)$ close to 1, the queue evolution is influenced more by the fluctuation than by the value of $1 - \rho(t)$ itself. Just a few steps would suffice to show the qualitative behavior.

Once $\lambda(t)$ has passed μ, the queue distribution will start to move slowly away from the boundary. If $\lambda(t)$ stays above μ long enough so that $Q(t)$ will almost certainly not vanish until $\lambda(t)$ is again less than μ, then it is very easy to follow the evolution of the queue distribution. Starting from an initial distribution evaluated at some time τ_n when the distribution of $Q(\tau_n)$ has moved away from the boundary, the changes in $Q(t)$ will be normally distributed, and independent of $Q(\tau_n)$ with a mean and variance given by

$$E\{Q(t) - Q(\tau_n)\} = \int_{\tau_n}^{t} [\lambda(t') - \mu]dt'$$

$$\mathrm{Var}\{Q(t) - Q(\tau_n)\} = \int_{\tau_n}^{t} [I_A\lambda(t') + I_D\mu]dt'. \qquad (9.2)$$

When, finally, the rush hour is nearly over and the queue distribution starts to hit the boundary again, the return to the quasi-equilibrium distribution can be described by approximations similar to (8.23). In essence, one simply takes the probability mass that would have crossed the boundary if the boundary were not there and redistributes it into the quasi-equilibrium distribution.

Obviously the explicit evaluation of the queue evolution in the manner suggested above would be quite tedious. The difficult part is to follow the queue distribution as $\rho(t)$ approaches or passes through 1. We will now try to examine this behavior in more detail by methods more amenable to numerical evaluation.

9.2 Small deviations from the equilibrium distribution

Although it is clear that (9.1) represents a first approximation for the queue distribution if $d\rho(t)/dt$ is sufficiently small, we might now ask if we can obtain a second approximation.

In the time-dependent diffusion equation

$$\frac{\partial F_Q(l;t)}{\partial t} = \left[(\mu - \lambda(t))\frac{\partial}{\partial l} + \frac{1}{2}(I_A\lambda(t) + I_D\mu)\frac{\partial^2}{\partial l^2} \right] F_Q(l;t)$$

we can immediately eliminate the time dependence of the variance term by using a nonlinear transformation of the time coordinate. If we define a new 'time' $t'(t)$ by

$$t'(t) = \int_0^t [I_A\lambda(\tau) + I_D]d\tau \tag{9.3}$$

and

$$F_Q'(l;t'(t)) = F_Q(l;t)$$

then

$$\frac{\partial F_Q(l;t)}{\partial t} = \left[I_A\lambda(t) + I_D\mu \right]\frac{\partial F_Q'(l;t')}{\partial t'}$$

and $F_Q'(l;t')$ satisfies the equation

$$\frac{\partial F_Q'(l;t')}{\partial t'} = \left[\xi(t')\frac{\partial}{\partial l} + \frac{1}{2}\frac{\partial^2}{\partial l^2} \right] F_Q'(l;t') \tag{9.4}$$

with

$$\xi(t') = \frac{1 - \lambda(t(t'))/\mu}{I_A\lambda(t(t'))/\mu + I_D}.$$

If $\xi(t') > 0$ during some time period, the quasi-equilibrium solution of (9.4) is

$$1 - F_Q'(l;t') = \exp(-2\xi(t')l)$$

with a mean of $1/2\xi(t')$. If we divide (9.4) by $\xi^2(t')$,

$$\frac{1}{\xi^2(t')}\frac{\partial F_Q'}{\partial t'} = \left[\frac{1}{\xi(t')}\frac{\partial}{\partial l} + \frac{1}{2\xi^2(t')}\frac{\partial^2}{\partial l^2} \right] F_Q'(l;t'),$$

we see that $1/\xi(t')$ acts like a unit of length, it is actually $L_0(t(t'))$, and $\xi^{-2}(t')$ acts like a unit of the time t'.

If we treat $\xi^{-1}(t')$ as a (time-dependent) unit of length, this suggests that we write

$$F'_Q(l;t') = F_Q^*(\xi(t')l;t') = F_Q^*(z;t'). \qquad (9.5)$$

Since

$$\frac{\partial F'_Q(l;t')}{\partial t'} = \frac{\partial F_Q^*(z;t')}{\partial t'} + l\frac{d\xi(t')}{dt'}\frac{\partial F_Q^*(z;t')}{\partial z},$$

$F_Q^*(z;t)$ satisfies the equation

$$\frac{1}{\xi^2(t')}\frac{\partial F_Q^*(z;t')}{\partial t'} = \left[(1+\gamma(t')z)\frac{\partial}{\partial z} + \frac{1}{2}\frac{\partial^2}{\partial z^2}\right]F_Q^*(z;t'), \qquad (9.6)$$

with

$$\gamma(t') = \frac{1}{2}\frac{d}{dt'}\xi^{-2}(t').$$

This is a convenient starting point for the analysis of small deviations from the quasi-equilibrium distribution. If $\xi(t')$ were constant, $\gamma(t')$ would be zero and (9.6) would have the stationary solution $\exp(-2z)$. The $\gamma(t')$ has been written in the above form because $\xi^{-2}(t')$ acts like a time unit and so $\gamma(t')$ represents a 'dimensionless' rate of change of this time unit. The distribution $F_Q^*(z;t')$ should stay close to the quasi-equilibrium distribution if $\gamma(t') \ll 1$ (which is essentially the condition (5.7)).

If it is assumed that the time scale $\xi^{-2}(t')$ is slowly varying in the sense that $\xi^{-2}(t')$ changes little in a 'time' unit $\xi^{-2}(t')$, it might be reasonable also to assume that it has a Taylor series expansion about some time t'_0.

$$\xi^{-2}(t') = \xi^{-2}(t'_0) + 2\gamma(t'_0)(t'-t'_0) + \frac{d\gamma(t'_0)}{dt'_0}(t'-t'_0)^2 + \ldots$$

and that, for $t'-t'_0$ comparable with $\xi^{-2}(t'_0)$, the higher order terms in the expansion are smaller than the linear term in $(t'-t'_0)$, (except possibly if $\xi^{-2}(t')$ has a maximum and $\gamma(t'_0) = 0$). Whereas in the first approximation we neglect γ completely, in the second approximation we might treat $\gamma(t')$ as (nearly) constant, and in subsequent approximations correct for the time dependence of $\gamma(t')$.

In (9.6) $\gamma(t')$ is multiplied by the dimensionless length z, a reflection of the fact that the larger queue lengths are slower to adjust to an equilibrium distribution than the short queues. Even though $\gamma \ll 1$, there will be values of z sufficiently large that γz is not small compared with 1. If, however, the queue distribution stays close to the equilibrium distribution, most relevant queue lengths should have z comparable with 1. If we treat γ as nearly constant, we should not take too seriously any possible consequences of $z \to \infty$ or $\gamma z \to \infty$. We seek solutions of (9.6) for which $1 - F_Q^*(z; t') \to 1$ for $z \to 0$ and behaves approximately (for $\gamma = 0$) like e^{-2z} for moderately large z.

If $\gamma(t') = \gamma$ were (exactly) constant, we might expect (9.6) to have a stationary solution with $\partial F_Q^*(z; t')/\partial t' = 0$ and $F_Q^*(z; t') = F_Q^*(z)$;

$$\left[(1 + \gamma z)\frac{d}{dz} + \frac{1}{2}\frac{d^2}{dz^2} \right][1 - F_Q^*(z)] = 0, \qquad (9.7)$$

If we were to solve this by perturbation methods, we would propose a solution of the type

$$1 - F_Q^*(z) = 1 - F_Q^{(0)}(z) - \gamma F_Q^{(1)}(z) \ldots$$

with

$$1 - F_Q^{(0)}(z) = e^{-2z},$$

and $F_Q^{(1)}(z)$ a solution of the equation

$$\left[\frac{d}{dz} + \frac{1}{2}\frac{d^2}{dz^2} \right] F_Q^{(1)}(z) = -2z\,e^{-2z},$$

with $F_Q^{(1)}(0) = 0, F_Q^{(1)}(\infty) = 0$. The solution in question is

$$F_Q^{(1)}(z) = (z + z^2)e^{-2z},$$

so to first order in γ,

$$1 - F_Q^*(z) \simeq e^{-2z}[1 - \gamma(z + z^2)]. \qquad (9.8)$$

Although this approximation is valid only for finite z and sufficiently small γ, it does display the unpleasant feature that for $\gamma > 0, 1 - F_Q^*(z)$ becomes negative for large z. Alternatively, we could solve (9.7) exactly. A first integration of (9.7) gives

$$\frac{d}{dz}[1 - F_Q^*(z)] = -C \exp(-2z - \gamma z^2).$$

This immediately presents a problem if $\gamma < 0$ because this derivative becomes infinite for $z \to \infty$. If, however, $\gamma > 0$ this can be integrated to give a solution satisfying the boundary conditions at $z = 0$ and ∞:

$$1 - F_Q^*(z) = \frac{\displaystyle\int_z^\infty \exp(-2y - \gamma y^2)dy}{\displaystyle\int_0^\infty \exp(-2y - \gamma y^2)dy}, \qquad \gamma \geq 0, \qquad (9.9)$$

This could also be expressed in terms of the Φ-functions, but for $\gamma \ll 1$ the above form is probably more convenient. If $\exp(-\gamma y^2)$ is approximated by $1 - \gamma y^2$ this will agree with (9.8) to first order in γ. If one prefers all approximations to be positive for all z then perhaps one should use (9.8) for $\gamma < 0$ and (9.9) for $\gamma > 0$.

If $\gamma(t') = \gamma > 0$ is exactly constant (but not necessarily small), (9.9) represents an exact solution of (9.6) starting from some appropriate initial conditions. That $\gamma(t') = \gamma$ implies

$$\xi^{-2}(t') = \left[\frac{I_A\lambda(t(t'))/\mu + I_D}{1 - \lambda(t(t'))/\mu}\right]^2 = \xi^{-2}(t'_0)[1 + 2\gamma\xi(t'_0)(t' - t'_0)],$$
(9.10)

i.e., the unit of time $\xi^{-2}(t')$ increases by approximately a factor $1 + 2\gamma$ in a time $\xi^{-2}(t'_0)$. Equivalently $1 - \rho$ is approximately proportional to $[1 + 2\gamma\xi_0(t'_0)(t' - t'_0)]^{-1/2}$. The traffic is increasing so that $\rho \to 1$ for $t' \to \infty$.

It is of no great practical importance in itself that there exists (for any $\gamma > 0$) a time-dependent arrival rate, i.e. $\xi^{-2}(t')$, such that the queue distribution depends upon time only through a scaling of the queue coordinate;

$$F'_Q(l;t') = F^*_Q(\xi(t')l)$$

with $F^*_Q(z)$ given by (9.9). The reason that there is no such exact solution for $\gamma < 0$ is that, for $\gamma < 0$, (9.10) would imply that ρ is decreasing from some proposed initial distribution at time t_0, which, for $t_0 \to -\infty$, would have $\rho = 1$, i.e., infinite queues.

Our conjecture is that if $\gamma(t')$ changes little in a time of order $\xi^{-2}(t')$, then (9.8) or (9.9) are appropriate solutions of (9.6) if γ is replaced by its value at time $t', \gamma(t')$.

We can always propose this as an approximate solution; let

$$1 - F^*_Q(z;t') = e^{-2z}[1 - \gamma(t')(z + z^2)] - F^{(2)}(z;t)$$
(9.11)

and obtain for $F^{(2)}(z;t)$ the equation

$$\left[(1 + \gamma(t')z)\frac{\partial}{\partial z} + \frac{1}{2}\frac{\partial^2}{\partial z^2}\right]F^{(2)}(z;t)$$

$$= -\gamma^2(t')e^{-2z}(z - 2z^3) + \frac{1}{\xi^2(t')}\frac{\partial F^{(2)}(z;t)}{\partial t}$$

$$+ \left[\frac{1}{\xi^2(t')}\frac{d\gamma(t')}{dt'}\right]e^{-2z}(z + z^2)$$

with the boundary conditions $F^{(2)}(0;t) = F^{(2)}(\infty;t') = 0$.

If we now neglect the $\gamma(t')z$ term on the left hand side and assume that the last term on the right hand side is nearly independent of t', we can evaluate a quasi-stationary solution of this equation, namely

$$F^{(2)}(z;t) \simeq -\gamma^2(t')(z + z^2 + z^3 + \frac{1}{2}z^4)e^{-2z}$$

$$-\frac{1}{2}\left[\frac{1}{\xi^2(t')}\frac{d\gamma(t')}{dt'}\right]\left(z + z^2 + \frac{1}{3}z^3\right)e^{-2z}. \qquad (9.12)$$

The main qualitative effect of a slowly varying arrival rate is that changes in the actual queue distribution lag behind the changes in the equilibrium distribution, particularly for the large queue lengths. This is already illustrated in the first correction (9.8). If $\gamma > 0$, the equilibrium queue is increasing, but the correction term reduces the probability in the tail of the distribution. Conversely, if $\gamma < 0$ the correction term places more probability in the tail. In (9.12) we see also that if $\gamma = 0$ but $d\gamma(t')/dt' < 0$ indicating that the equilibrium queue length has a maximum, the correction to $1 - F_Q^*(z;t)$ is negative; the queue is shorter than the equilibrium queue, as it was at an earlier time.

From (9.11), (9.12) one can evaluate various moments of Q, for example

$$E\{Q(t)\} \simeq \frac{1}{2\xi(t')}\left[1 - \gamma(t') + \frac{5}{2}\gamma^2(t') + \frac{5}{8\xi^2(t')}\frac{d\gamma(t')}{dt'}\right], \qquad (9.13)$$

$$E\{Q^2(t)\} \simeq \frac{1}{2\xi^2(t')}\left[1 - \frac{5}{2}\gamma(t') + \frac{37}{4}\gamma^2(t') + \frac{7}{4\xi^2(t')}\frac{d\gamma(t')}{dt'}\right]. \qquad (9.14)$$

A more intuitive form of this, however, can be obtained if we observe that

$$\gamma(t') = \frac{1}{2}\frac{d}{dt'}\left(\frac{1}{\xi^2(t')}\right) = \frac{1}{\xi(t')}\frac{d}{dt'}\left(\frac{1}{\xi(t')}\right)$$

and that

$$\frac{1}{\xi(t' - 1/\xi^2(t'))} = \frac{1}{\xi(t')} - \frac{1}{\xi^2(t')}\frac{d}{dt'}\left(\frac{1}{\xi(t')}\right)$$

$$+ \frac{1}{2}\frac{1}{\xi^2(t')}\frac{d^2}{dt'^2}\left(\frac{1}{\xi(t')}\right) + \cdots$$

To the same order of approximation, we can also write (9.13) as

$$E\{Q(t)\} = \frac{1}{2\xi(t' - \xi^{-2}(t'))}$$

$$\times \left[1 + \frac{25}{8\xi^2(t')} \left(\frac{d}{dt'} \frac{1}{\xi(t')} \right)^2 + \frac{1}{8\xi^3(t')} \left(\frac{d^2}{dt'^2} \frac{1}{\xi(t')} \right) \right]$$

(9.15)

and correspondingly

$$E\{Q^2(t)\} = \frac{1}{2\xi^2(t - 5\xi^{-2}(t')/4)}$$

$$\times \left[1 + \frac{37}{16} \left(\frac{d}{dt'} \frac{1}{\xi^2(t')} \right)^2 + \frac{3}{32\xi^2(t')} \left(\frac{d^2}{dt'^2} \frac{1}{\xi^2(t')} \right) \right].$$

(9.16)

The second and third terms in the brackets are 'second order' effects. The first order effect of a slowly varying $\rho(t)$ is that the mean queue is approximately equal to the equilibrium mean queue at an earlier time, $t' - \xi^{-2}(t')$, and the mean square length is approximately equal to its equilibrium value at time $t' - 5\xi^{-2}(t')/4$. The higher moments being more sensitive to the tail of the queue distribution tend to lag more behind their equilibrium values.

Usually, if the second order terms in (9.15), (9.16) are not negligibly small, the whole approximation scheme is about to collapse if $\rho(t)$ should increase any more. We are mainly interested in these terms in the special case in which $\rho(t)$ reaches a maximum $\rho(t_0) < 1$ at some time t_0, and $1 - \rho(t_0)$ is sufficiently large that the above approximations are valid. The first approximation predicts that $E\{Q(t)\}$ has a maximum at $t' = t_0' + \xi^{-2}(t_0')$ (nearly) equal to the maximum equilibrium queue length $(2\xi(t_0'))^{-1}$ at t_0'. The second term in the bracket of (9.15) is positive but also very small if $\xi^{-1}(t')$ is near its maximum where its derivative vanishes. The third term in the bracket is negative near t_0'; consequently, the maximum of $E\{Q(t)\}$ is (slightly) less than the maximum of $[2\xi(t')]^{-1}$. A similar type behavior applies also to $E\{Q^2(t)\}$.

9.3 Transition through saturation

If $\rho(t)$ should increase and pass through 1, we may choose the time origin to be that time when $\rho(t)$ exceeds 1 for the first time. The approximations of the last section will certainly fail as $\rho(t)$ approaches 1 and the unit of length $\xi^{-1}(t)$ becomes infinite.

The mean queue length $E\{Q(t)\}$ at $t = 0$ will obviously depend upon the rate at which $\rho(t)$ approaches 1. At one extreme if $\rho(t) \ll 1$ for $t < 0$ and suddenly jumps to a value greater than 1 for $t > 0$, then $E\{Q(0)\} \ll 1$ since a queue cannot form instantaneously with a finite arrival rate. At the other extreme, if $\rho(t)$ approaches 1 arbitrarily slowly the queue distribution will stay close to the quasi-equilibrium distribution until t is very close to 0 by which time $E\{Q(t)\}$ will be very large and still growing.

If $\rho(t)$ passes through 1, there is no reason why it should not behave smoothly there. Since in (9.4) $\xi(t')$ will vanish at $t' = 0$, suppose that over some relevant time period $\xi(t')$ can be approximated by a linear function

$$\xi(t') = -\alpha t', \quad \alpha > 0. \tag{9.17}$$

If we rescale the time and length coordinates so that

$$l^* = l/L_1, \quad t^* = t'/T_1' = t/T_1$$

(9.4) becomes

$$\frac{\partial F_{Q^*}(l^*; t^*)}{\partial t^*} = \left[-\frac{T_1'^2}{L_1} \alpha t^* \frac{\partial}{\partial l^*} + \frac{1}{2} \frac{T_1'}{L_1^2} \frac{\partial^2}{\partial l^{*2}} \right] F_{Q^*}(l^*; t^*).$$

In particular, if we choose

$$\frac{T_1'^2}{L_1} \alpha = \frac{T_1'}{L_1^2} = 1$$

i.e.,

$$T_1' = L_1^2 = \alpha^{-2/3}, \tag{9.18}$$

then

$$\frac{\partial F_{Q^*}(l^*; t^*)}{\partial t^*} = \left[-t^* \frac{\partial}{\partial l^*} + \frac{1}{2} \frac{\partial^2}{\partial l^{*2}} \right] F_{Q^*}(l^*; t^*). \tag{9.19}$$

This must satisfy the usual boundary conditions $F_{Q^*}(0; t^*) = 0$, $F_{Q^*}(\infty; t^*) = 1$. In addition, if the linear approximation (9.17) is valid for $t^* < 0$ and $|t^*| \gg 1$, we seek the solution of (9.19) which came from a quasi-equilibrium distribution.

$$F_{Q^*}(l^*; t^*) \simeq 1 - \exp(+2t^* l^*) \quad \text{for } -t^* \gg 1.$$

There is no simple analytic solution of (9.19), but this equation contains no parameters. There is only one universal solution which, except for a scale of coordinates, describes the evolution of the queue distribution for any system with a linearly increasing $\xi(t')$. Conversely, the evolution of any $F_Q(l; t)$ for which (9.17) holds, evaluated by simulation, numerical integration or whatever, can be scaled into a solution of (9.19). The main qualitative features of the solution, however, are already contained in the scaling factors. The solution of (9.19) obviously must imply that the $Q^*(t^*)$ is comparable with 1 when $|t^*|$ is comparable with 1 since the only numbers in (9.19) are 1 or 1/2. The equation must also describe a queue distribution which for $-t^* \gg 1$ is nearly in equilibrium but for $t^* \gg 1$ is approximately normal.

In terms of the original units, this means that $Q(0)$ must be comparable with L_1 when $|t|$ is comparable with T_1. From (9.3) we can relate T_1 to T'_1:

$$T_1 \simeq \frac{T'_1}{(I_A + I_D)\mu} = \frac{\alpha^{-2/3}}{(I_A + I_D)\mu} \tag{9.20}$$

or we can relate the time T_1 to the value of $1 - \rho(-T_1)$ from (9.4),

$$1 - \rho(-T_1) \simeq (I_A + I_D)\alpha T'_1 = (I_A + I_D)\alpha^{+1/3}. \tag{9.21}$$

From (9.18)

$$L_1 = \alpha^{-1/3}. \tag{9.22}$$

The α can also be related to $d\rho(t)/dt$,

$$\alpha = \frac{d\xi(t')}{dt'}\bigg|_{t'=0} = -\frac{d}{dt'}\frac{1 - \rho(t(t'))}{I_A\rho(t(t')) + I_D}\bigg|_{t'=0} = \frac{1}{(I_A + I_D)^2}\frac{d\rho(t)}{\mu\,dt}\bigg|_{t=0}.$$

$$\tag{9.23}$$

The factor $\mu^{-1}d\rho(t)/dt$ in (9.23) can be interpreted as the change in $\rho(t)$ during an average service time which in any applications of diffusion approximations should be numerically quite small. In typical 'rush hour' situations we might expect $\rho(t)$ to increase at a rate comparable with one per hour so that α^{-1} is comparable with the hourly service rate. For a busy highway α^{-1} might be in the range 10^3 to 10^4 but for other facilities with service times measured in minutes α^{-1} would typically be of order 10^2.

Although α itself is usually 'very small', the key parameter in these formulas is $\alpha^{1/3}$, which may not be very small. For the highway example with $\alpha \simeq 10^{-3}$ or 10^{-4}, $\alpha^{1/3} \simeq 1/10$ or $1/20$ which means that the transition region is for $|1 - \rho| \lesssim 1/10$ or $1/20$ and the typical queue length is about 10 to 20. For $\alpha \simeq 10^{-2}$, the transition region is for $|1 - \rho| \lesssim 1/5$ and the typical queue length is about 5.

Approximate numerical solutions of (9.19) have been evaluated. The simplest procedure is to follow the evolution of the $p_j(t)$ for a random walk queue such as described in Section 5.3, but with λ replaced by a $\lambda(t)$ which increases linearly with t at a very slow rate. The resulting queue distribution can then be rescaled so as to be an approximate solution of (9.19). The slower the rate of growth of λ, the closer the discrete distribution is to the solution of (9.19).

Figs 9.1 and 9.2 show some properties of the queue distribution. Fig. 9.1 shows $E\{Q^*(t^*)\}$ i.e., $E\{Q(t')\}/L_1$ as a function of $t^* = t'/T'_1$. The broken line curve for $t^* < 0$ is the quasi-equilibrium mean queue $-1/(2t^*)$. As t^* increases toward zero $E\{Q^*(t^*)\}$ remains finite and approaches a value of about 0.65 while $-1/(2t^*)$ becomes infinite. The broken line curve for $t^* > 0$, $t^{*2}/2$, represents the deterministic fluid approximation. Once the queue has become sufficiently large that there is a negligible probability for the queue to vanish at any later time, the continuum approximation will correctly describe the future growth of the mean queue, i.e., $E\{Q^*(t^*)\} - t^{*2}/2$ should approach a constant for $t^* \gg 1$. Indeed, it has almost reached its limiting value of about 0.95 for $t^* = 1$.

Fig. 9.1 gives an approximate description of the behavior of

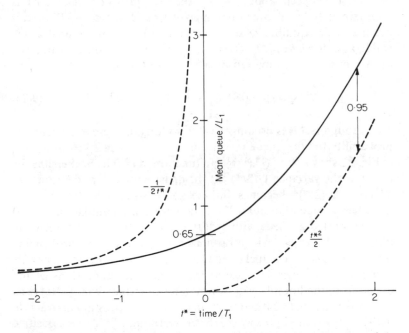

Figure 9.1 *Evolution of the mean queue length as $\rho(t)$ passes through 1*

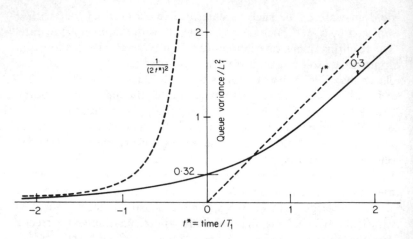

Figure 9.2   *Evolution of the queue variance as $\rho(t)$ passes through 1*

$E\{Q^*(t^*)\}$ or $E\{Q(t)\}$ over the range of t^* from about -2 to $+1$, i.e., for t between about $-2T_1$ and T_1, provided that $\xi(t')$ is approximately linear over this range of time. For $t \leq -2T_1$ we can use the quasi-equilibrium solutions (regardless of whether (9.17) is valid) and for $t \geq T_1$, $E\{Q(t)\}$ will continue to differ from the deterministic approximation by about $0.95\,L_1$, i.e.,

$$E\{Q(t)\} \simeq 0.95\,L_1 + \int_0^t [\lambda(t') - \mu]\,dt' \qquad (9.24)$$

even though (9.17) is no longer valid, so long as there is a negligible probability for the queue to vanish.

Fig. 9.2 shows $\mathrm{Var}\{Q^*(t^*)\}$ as a function of t^*. The broken line for $t^* < 0$ is the variance $1/(2t^*)^2$ of the quasi-equilibrium distribution. Whereas $1/(2t^*)^2$ becomes infinite for $t^* \to 0$, $\mathrm{Var}\{Q^*(t^*)\}$ approaches a value of about 0.32. If there were no boundary at $l^* = 0$ and we started a queue with $Q^*(0) = 0$, $Q^*(t^*)$ would be normally distributed for $t^* > 0$ with a variance of t^*, shown by the broken line for $t^* > 0$. Once the actual queue has escaped from the boundary, its variance should also increase at this same rate. Reflections of the queue from the boundary, however, tend to reduce $\mathrm{Var}\{Q^*(t^*)\}$ as was shown also in (8.24) and (8.25) in the case of a constant arrival rate with $\rho > 0$. From Fig. 9.2 we see that, even though $Q^*(0)$ has a positive variance, subsequent reflections from the boundary more than compensate for this so as to give $\mathrm{Var}\{Q^*(t)\} < t^*$ for $t^* \gtrsim 1/2$ and $\mathrm{Var}\{Q^*(t^*)\} - t \simeq -0.3$ for $t^* \gtrsim 1.5$.

Even if the linear approximation should fail for $t^* \gtrsim 1.5$, it would still be true that

$$\mathrm{Var}\{Q(t)\} \simeq -0.3\,L_1^2 + t'$$
$$= -0.3L_1^2 + \int_0^t [I_A\lambda(t') + I_D\mu]dt' \qquad (9.25)$$

for $t^* \gtrsim 1.5$ so long as $Q(t)$ is unlikely to vanish. Furthermore, the distribution of $Q(t)$ will rapidly approach a normal distribution for $t^* \gtrsim 2$ with a mean and variance given by (9.24) and (9.25).

9.4 A mild rush hour

The results of the last section, in essence, describe a stochastic correction for the graphical deterministic solution of Fig. 2.3, represented analytically by the second term of (9.24), provided that the queue distribution escapes from the boundary. If the queue distribution becomes approximately normal by the time $\lambda(t)$ has passed its maximum and dropped below μ again, one can also easily approximate the evolution of the queue distribution as it approaches the boundary again and redistributes itself into a quasi-equilibrium distribution. The 'end of the rush hour' occurs, on the average, somewhat later than predicted by the deterministic approximation because of the extra term in (9.24), and is also smeared over a period of time.

Suppose, however, that $\lambda(t)$ reaches a maximum such that a significant queue exists only for a short period of time during which $\lambda(t)$ can be approximated by a quadratic function as in Section 2.3.

$$\lambda(t) = \lambda(t_1) - \beta(t - t_1).$$

If $\lambda(t_1)$ exceeds μ for a sufficiently long time we can still apply the results of the last section, or if $\lambda(t_1)$ is less than μ and changing at a sufficiently slow rate the quasi-equilibrium results of Section 9.2 can be applied. There is a range of $|\lambda(t_1) - \mu|$, however, in which neither approximation applies. For $\lambda(t_1) > \mu$ the stochastic correction to the queue dominates the deterministic approximation of Section 2.3 but, for $\lambda(t_1) < \mu$, $\lambda(t)$ is changing too rapidly for the quasi-equilibrium behavior to apply.

We can again eliminate the time dependence of the variance terms in the diffusion and start from (9.4). To assume that $\lambda(t)$ is approximately a quadratic function of t is nearly equivalent to assuming that $\xi(t')$ can be approximated by a quadratic function of t'. The latter is more convenient, however. Suppose that

$$\xi(t') = \xi(t'_1) + \beta'(t' - t'_1)^2, \quad \beta' > 0. \tag{9.26}$$

If we choose scales L_2, T'_2 for l, t',

$$l^* = l/L_2, \quad t^* = t'/T'_2$$

the diffusion equation takes the form

$$\frac{\partial F_{Q^*}(l^*; t^*)}{\partial t^*} = \left\{ \left[\frac{\xi(t'_1)T'_2}{L_2} + \frac{\beta' T'^3_2}{L_2} t^{*2} \right] \frac{\partial}{\partial l^*} + \frac{1}{2} \frac{T'_2}{L^2_2} \frac{\partial^2}{\partial l^{*2}} \right\}$$
$$\times F_{Q^*}(l^*; t^*).$$

In particular, if we choose

$$\frac{T'_2}{L^2_2} = \frac{\beta' T'^3_2}{L_2} = 1,$$

i.e.,

$$T'_2 = \beta'^{-2/5}, \quad L_2 = \beta'^{-1/5} \tag{9.27}$$

the diffusion equation becomes

$$\frac{\partial F_{Q^*}(l^*; t^*)}{\partial t^*} = \left[(\varepsilon + t^{*2}) \frac{\partial}{\partial l^*} + \frac{1}{2} \frac{\partial^2}{\partial l^{*2}} \right] F_{Q^*}(l^*; t^*), \tag{9.28}$$

with

$$\varepsilon = \xi(t'_1)L_2 = \xi(t')\beta'^{-1/5}.$$

Since (9.28) contains only the parameters ε and 1, it is clear that the qualitative behavior of the solution will depend on whether $\varepsilon \gg 1$, $|\varepsilon|$ is comparable with 1, or $-\varepsilon \gg 1$. For $\varepsilon \gg 1$ the system will be undersaturated and the queue distribution should behave approximately like the quasi-equilibrium distribution. If $-\varepsilon \gg 1$ the system is oversaturated sufficiently that the behavior of Section 9.3 applies.

For $|\varepsilon|$ comparable with 1 an accurate solution of (9.28) can be obtained only by some numerical integration but the most important qualitative features of the queue behavior are contained already in the scaling factors T'_2 and L_2 since the solution of (9.28) certainly must imply that $Q^*(t^*)$ is comparable with 1 for $|t^*|$ comparable with 1.

In terms of the original units, for $\rho(t_1)$ close to 1,

$$\xi(t'_1) = \frac{1 - \rho(t_1)}{I_A \rho(t_1) + I_D} \simeq \frac{1 - \rho(t_1)}{I_A + I_D},$$

and

$$\beta' \simeq \left[-\frac{1}{2\mu^2} \frac{d^2\rho(t_1)}{dt_1^2} \right] \frac{1}{(I_A + I_D)^3},$$

$$\varepsilon = \frac{[1 - \rho(t_1)]}{(I_A + I_D)^{2/5} \left[-\frac{1}{2\mu^2} \frac{d^2\rho(t_1)}{dt_1^2} \right]^{1/5}}, \tag{9.29a}$$

$$T_2 = (I_A + I_D)^{1/5} \left[-\frac{1}{2\mu^2} \frac{d^2\rho(t_1)}{dt_1^2} \right]^{-2/5}, \tag{9.29b}$$

$$L_2 = (I_A + I_D)^{3/5} \left[-\frac{1}{2\mu^2} \frac{d^2\rho(t_1)}{dt_1^2} \right]^{-1/5}. \tag{9.29c}$$

In (9.29a), $\mu^{-2}d^2\rho(t_1)/dt_1^2$ represents the second derivative of $\rho(t_1)$ with respect to a time measured in units of the service time $1/\mu$. In typical applications we expect β' to be numerically very small compared with 1. If, for example, $\rho(t)$ should decrease to nearly zero for $|t - t_1|$ approximately one hour, β' would be comparable with the square of the service time measured in hours. For highway traffic with service times of a few seconds, β' would typically be of order 10^{-7}. The key parameter in (9.29a) or (c), however, is $\beta'^{-1/5}$.

For $\beta' \simeq 10^{-7}$, L_2 is only about 30, but the range $|\varepsilon| \lesssim 1$ applies only for $|1 - \rho(t_1)| \lesssim 1/30$, i.e., $0.97 \lesssim \rho(t_1) \lesssim 1.03$. This narrow range for $\rho(t_1)$ is not likely to arise very often in applications. For service facilities with service times comparable with a minute, however, $\beta' \simeq 10^{-3}$ or 10^{-4}; L_2 is in the range of about 5 to 10 and $|\varepsilon| \le 1$ means $|1 - \rho| \le 0.1$ or 0.2. Queues of this magnitude for aircraft landing on a runway would be considered as 'quite congested'. Although $\rho(t)$ does exceed 1 at some airports, a maximum $\rho(t)$ of about 0.85 is sufficient to cause concern by airport planners.

Approximate numerical solutions of (9.28) were obtained by evaluating the $p_j(t)$ for a random walk queue with time-dependent $\lambda(t)$, as for (9.19). Fig. 9.3 shows $E\{Q^*(t^*)\}$ as a function of t^* for several values of ε. The curve for $\varepsilon = 2$ is barely distinguishable from a curve (not shown) for the quasi-equilibrium mean queue $[2(\varepsilon + t^2)]^{-1}$. The main difference is that the maximum of the former curve is delayed slightly. To within the accuracy of the numerical solution the displacement of the maximum is consistent with the value of $\xi^{-2}(t_1) = 1/4$ predicted by (9.15).

The curve for $\varepsilon = 1$ deviates appreciably from the quasi-equilibrium mean shown by the broken line curve. This ε is already

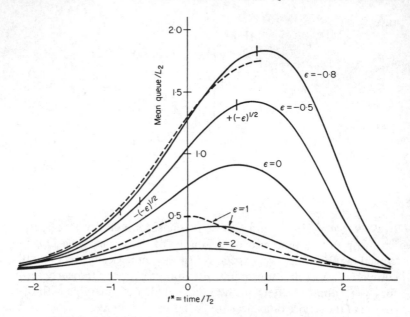

Figure 9.3   *Evolution of the mean queue for a mild rush hour*

too small for (9.15) to apply but the curve does show the type of qualitative effects predicted by (9.15). The maximum mean queue is delayed and of smaller amplitude than the equilibrium mean. For $\varepsilon = 0$, $E\{Q^*(t^*)\}$ reaches a maximum of about 0.92 at $t^* \simeq 0.65$.

For $\varepsilon = -0.5$ and $\varepsilon = -0.8$, $\rho(t_1) > 1$ and $\rho(t^*)$ passes through 1 at $t^* = \pm(-\varepsilon)^{1/2}$ identified on these graphs by the vertical markers on the curves. For $\varepsilon = -0.8$ the curve for $E\{Q^*(t^*)\}$ is already following quite closely the type of behavior described in Section 9.3 for the transition near $t^* = -(-\varepsilon)^{1/2}$. The mean queue reaches a maximum for $t^* \simeq +(-\varepsilon)^{1/2}$ as predicted by the fluid approxima- tion (and the correction (9.24) to the fluid approximation). The broken line curve near $\varepsilon = -0.8$ is an approximate curve obtained from the results of Section 9.3. For still more negative ε the approximations of Section 9.3 will be quite accurate.

Fig. 9.4 shows similar curves for $\text{Var}\{Q^*(t)\}$. For $\varepsilon = +2$, the variance is nearly symmetric about $t^* = 0$ but delayed somewhat, more so than $E\{Q^*(t^*)\}$. These curves all show the same qualitative behavior as $E\{Q^*(t^*)\}$ but all effects are exaggerated because the variance is more sensitive to the tail of the queue distribution which changes at a slower rate than the short queues.

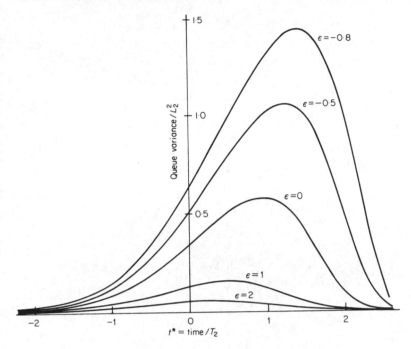

Figure 9.4   *Evolution of the queue variance for a mild rush hour*

Fig. 9.5 shows the maximum expected queue $\max_t E\{Q^*(t^*)\}$ as a function of ε. The circled points are the results of the numerical calculations shown in Fig. 9.3. The 'deterministic theory' is taken from (2.5). The curve $|\varepsilon| \gg 1$, $\varepsilon < 0$ includes the stochastic correction (9.24) and the curve labeled $|\varepsilon| \gg 1$, $\varepsilon > 0$ is the quasi-equilibrium approximation. The curve describing the correction (9.15) to the maximum quasi-equilibrium mean queue has not been drawn because, over the range of ε where this formula would apply, the error in the numerical solution is much larger than the correction term (which is proportional to ε^{-5}).

These graphs for $|\varepsilon|$ comparable with 1 show that most properties of the queue distribution are, in a sense, 'smooth interpolations' between what happens for $\varepsilon \gtrsim 1$ and $\varepsilon \lesssim 1$. This is to be expected because, if $\rho(t)$ has a maximum close to 1, the size of the queue will be most sensitive to fluctuations in the arrivals (or departures) which cause an excess of total arrivals (or deficiency of departures) over the finite time when queueing is most likely to occur. The system, of

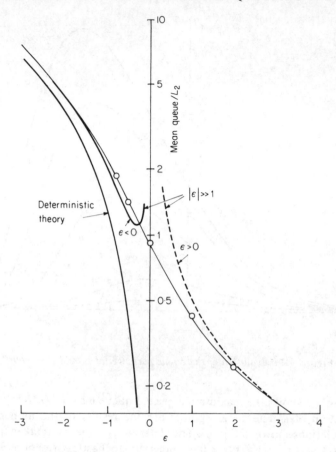

Figure 9.5 *The maximum mean queue during a rush hour*

course, responds to the actual arrivals and departures (not their expectations) which over any finite time period could appear as if they were generated from a 'randomly selected $\rho(t)$'. Qualitatively, the queue behaves like some appropriate average of the behaviors for undersaturated and oversaturated conditions.

9.5 Pulsed service, queue clears

In Section 9.1 it was pointed out that one could follow the evolution of the queue distribution for a slowly varying $\rho(t)$ by approximating $\rho(t)$ with a piecewise constant function and using the transient solution for the behavior of the queue distribution over successive periods of constant ρ. There are many important types of service systems, however, which behave almost as if $\rho(t)$ were piecewise constant but with very substantial jumps at various discrete times. Examples of such systems with steady arrivals but pulsed service were analyzed previously in Section 2.6 by deterministic approximations. There is also an analogous class of systems for which the arrivals occur in pulses (batches) but the service is steady.

In Section 2.6 it was assumed that the (deterministic) queue vanished during each service pulse but the duration of the service pulse and the time between pulses were allowed to vary from cycle to cycle. Although no specific assumptions were made regarding any stochastic behavior, there was certainly an implication that the red and green times R and G of a traffic signal, for example, could be treated as random variables having a distribution function determined by the frequency of occurrence of various values of r_i and g_i over a long time period. It was shown, in particular, that with irregular (random) cycle times both the mean and the variance of the cycle time affected the average delay or queue length.

We wish now to extend the theory described in Section 2.6 to include effects caused by a stochastic arrival and departure process in addition to a possible stochastic pattern of times R and G. Even if the queue vanishes during each service pulse we would like to estimate any extra total wait attributed to the fluctuations in arrivals or service (above those associated with the deterministic approximations). It is also of some interest to know the distribution of the total number of customers served and/or the duration of the busy period in each cycle. In the application to a public transportation system (buses), for example, with vehicles of finite capacity M, the distribution of the number of customers wishing to be served (board a bus) in one cycle given that the queue cleared in the previous cycle gives a measure of possible overloading of the vehicle or customers who could not be served by that vehicle. In the application to a traffic signal the distribution of the busy period will give a measure of the probability that the queue will clear in the next cycle, given that it cleared in the previous one. For some systems, such as an elevator, the number of customers who board at the lobby in one cycle (are served) also affects the trip time, i.e., the time of the next service pulse (loading at the lobby).

It was implied above that the queue might not vanish every cycle even though we expect it to do so most of the time. There are also situations (particularly for traffic signals) in which a queue will not clear for many cycles and may even become several times as large as can be served in one cycle. Here one might be interested particularly in the stochastic behavior of the overflow queue, the queue left after each service pulse. In the queueing literature a system which serves a finite number (larger than 1) of customers (almost) simultaneously is called a bulk server.

We consider first the behavior of $Q(t)$ starting at a time $t = 0$ when the service has just stopped and $Q(0) = 0$. Suppose that there is no service for a (random) time R during which customers arrive at a (nearly) constant rate λ and variance rate $I_A \lambda$. At time R the arrival process continues but the service commences at a rate $\mu > \lambda$ and variance rate $I_D \mu$.

In most of the intended applications to moderately congested systems the expected number of arrivals λR during a given time R is likely to be comparable with 10 or more in which case it would be reasonable to assume that $Q(R|R = t)$ is approximately normally distributed with mean λt and variance $I_A \lambda t$. In applications to buses, elevators, etc., the service rate μ (loading rate) is typically large compared with λ. The issue is not the time it takes to load but whether or not $Q(t)$ exceeds some limit M, the capacity of the vehicle. If R were deterministic (regularly spaced headways) the condition that there be only a small probability that $Q(R) > M$ is that $M - \lambda R$ be larger than one or two standard deviations $(I_A \lambda R)^{1/2}$, i.e.,

$$P\{Q(R) > M\} \simeq 1 - \Phi\left[\frac{M - \lambda R}{(I_A \lambda R)^{1/2}}\right] \ll 1. \qquad (9.30)$$

Until λ is sufficiently large that (9.30) is violated, the deterministic approximations of Section 2.6 actually describe $E\{Q(t)\}$. Since for $\lambda R > M$ the system is oversaturated, there is only a relatively small range of λ in which stochastic queueing effects are important, namely, for

$$1 - \frac{1}{(I_A M)^{1/2}} \lesssim \lambda R / M = \rho < 1.$$

(For a 16-passenger elevator $0.75 \lesssim \rho < 1$, for a 50-passenger bus $0.85 \lesssim \rho < 1$.)

If, on the other hand, R is random with a probability density $f_R(t)$, the (unconditional) distribution of $Q(R)$ gives

$$P\{Q(R) > M\} = \int_0^\infty \left[1 - \Phi\left(\frac{M - \lambda t}{(I_A \lambda t)^{1/2}}\right)\right] f_R(t)\, dt. \qquad (9.31)$$

This distribution is typically far from normal and has moments

$$E\{Q(R)\} = \lambda E\{R\}$$
$$\text{Var}\{Q(R)\} = E\{\text{Var}\{Q(R)|R\}\} + \text{Var}\{E\{Q(R)|R\}\}$$
$$= \lambda E\{R\}[I_A + \lambda E\{R\}C^2(R)]. \qquad (9.32)$$

Unless $C^2(R)$ is quite small, less than $1/\lambda E\{R\} \simeq 1/M$, $\text{Var}\{Q(R)\}$ is likely to be determined mostly by the fluctuations in R rather than the arrival process. If this is so, the distribution of $Q(R)$ will also be determined essentially by the distribution of R,

$$P\{Q(R) > M\} \simeq P(R > M/\lambda).$$

Queueing effects will be important for a much wider range of λ than for regularly spaced headways. This would be the typical situation for a busy bus route having mean headways less than 10 minutes. Since it is difficult to control short headways (except possibly at the dispatch point), $C^2(R)$ is likely to be comparable with 1.

If it is still true, even with a random R, that $P\{Q(R) > M\} \ll 1$, expected queue lengths, waiting times, etc., will be as described in Section 2.6. Fluctuations in R will contribute to the average wait even though the queue empties during each service pulse.

For traffic signals μ is not generally large compared with λ. Although there may be a constraint M on the size of queue that can be stored (a finite waiting room), the more common concern is whether or not the queue clears during the green time and, if so, when. If the queue $Q(R)|R$ for given R is normally distributed, the evolution of the queue during the time R to $R + G$ is very similar to the transient behavior of $Q(t)$ discussed in Section 8.3. In models of traffic signals, however, it is customary to assume that, if the queue vanishes during the green time, it will not reform during the remaining green. The argument is that if any cars approach a signal with close spacing after the queue has cleared, they will probably also leave with the same spacing (not with a mean headway $1/\mu$). A reasonable model for the queue evolution, therefore, would be one for which $Q(t) = 0$ is treated as an absorbing boundary rather than a reflecting boundary (as a practical matter, however, it does not make much difference).

Until the queue is likely to vanish $Q(t)|R$ is approximately normal with

$$\text{Var}\{Q(t)|R\} = I_A \lambda t + I_D \mu(t - R)$$
$$= (I_A \lambda + I_D \mu)\left[t - \frac{I_D \mu R}{I_A \lambda + I_D \mu}\right]$$

and

$$E\{Q(t)|R\} = \lambda R - (\mu - \lambda)(t - R)$$
$$= -(\mu - \lambda)\left[t - \frac{I_D\mu R}{I_A + I_D\mu}\right] + \frac{\mu R(I_A + I_D)\lambda}{I_A\lambda + I_D\mu}. \quad (9.33)$$

The variance, being a linear function of t for $t > R$, behaves as if the variance had been increasing at a rate $I_A\lambda + I_D\mu$ since a time t_0

$$t_0 = \left(\frac{I_D\mu}{I_A\lambda + I_D\mu}\right)R < R, \quad (9.34)$$

when $\text{Var}\{Q(t_0)|R\} = 0$. The mean queue which decreases linearly with t behaves as if it had a value

$$Q(t_0)|R = \frac{\lambda R(I_A + I_D)\mu}{I_A\lambda + I_D\mu} \quad (9.35)$$

at time t_0. Thus the evolution of $Q(t)|R$ for $t > R$ is the same as for a hypothetical process that started at $Q(t_0)|R$ at time t_0 as illustrated in Fig. 9.6.

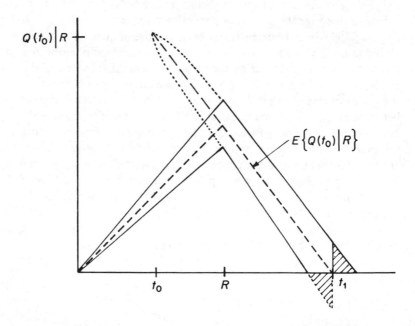

Figure 9.6   *Realizations of $Q(t)|R$ for a traffic signal*

The solution of the diffusion equation for the probability density with an absorbing barrier is obtained from a slight modification of (8.21),

$$
f_Q(l;t \mid R) = \frac{1}{[2\pi(I_A\lambda + I_D\mu)(t - t_0)]^{1/2}}
$$

$$
\left\{ \exp\left[-\frac{(l - Q(t_0) \mid R + (\mu - \lambda)(t - t_0))^2}{2(I_A\lambda + I_{D.}\mu)(t - t_0)} \right] \right.
$$

$$
-\exp\left[-\frac{2(\mu - \lambda)l}{(I_A\lambda + I_D\mu)} \right]
$$

$$
\left. \times \exp\left[\frac{-(l + Q(t_0) \mid R - (\mu - \lambda)(t - t_0))^2}{2(I_A\lambda + I_D\mu)(t - t_0)} \right] \right\}. \qquad (9.36)
$$

The second term of (9.36) describes the phenomenon that if $Q(t)$ comes close enough to the boundary the fluctuations in $Q(t)$ will cause it to be absorbed even before it would reach the boundary due to the mean drift. A typical value of λ/μ would be about $1/2$ or less, however, and the exponential factor on this term describing the equilibrium queue distribution has a mean of less than about $1/2$. Given that the actual queue length is integer valued, the accuracy of this term is questionable but also not very important. In essence, (9.36) describes a distribution which is moving toward the boundary at a rate $\mu - \lambda$ and also spreading. The distribution at any t is almost a truncated normal distribution but with a little extra probability absorbed near the boundary.

Even if the green time were sufficiently long that the queue was (almost) certain to vanish, fluctuation in the arrivals and departures cause the average delay and the mean queue to be slightly larger than predicted by the deterministic approximation of Section 2.6. In the deterministic approximation the total delay in one cycle is the area under the broken line curve of Fig. 9.6, which decreases at a rate $\mu - \lambda$ for $t > R$, but the expected total delay is the area under a realization of $Q(t)$ averaged over all realizations.

One can obtain a simple estimate of the extra delay caused by fluctuations by noting that if there were no boundary (the queue could be negative) then $E\{Q(t) \mid R\}$ would be given exactly by the broken line curve of Fig. 9.6 for $t < t_1$. If some queue realization (lower solid line) should vanish before t_1, we would replace the negative queue by zero thereby adding to the total delay a contribution equal to the shaded area of Fig. 9.6. On the other hand, if some realization of $Q(t)$ is positive at time t_1, there will be a delay after time t_1 equal to the shaded area of Fig. 9.6 for $t > t_1$. In either case the area of the triangular shaped region is approximately

$$\frac{1}{2}[Q(t_1)|R]^2/(\mu - \lambda),$$

with $Q(t_1)|R$ the queue that would exist without the boundary. The average excess delay due to fluctuations is therfore

$$\frac{1}{2}\text{Var}\{Q(t_1|R\}/(\mu - \lambda) = \frac{\lambda R(I_A + I_D)\mu}{2(\mu - \lambda)^2}.$$

If we add this to the deterministic part of the total delay, we have

$$\text{total delay}|R \simeq \frac{1}{2}\frac{R^2\lambda}{(1 - \lambda/\mu)}\left[1 + \frac{(I_A + I_D)}{R(\mu - \lambda)}\right]. \qquad (9.37)$$

Since in typical applications we expect $R(\mu - \lambda)$ to be moderately large (more than 5) this correction term is not very important. A more accurate evaluation of the correction obtained directly from (9.36) would also include some terms of order $[R(\mu - \lambda)^{-3/2}]$ or smaller.

If there is a sequence of cycles with red times r_i as in Section 2.6 but the queue clears (almost) every cycle, the total delay over many cycles can be obtained by summing (9.37) over all r_i or equivalently by taking the expectation of (9.37) with respect to a distribution of R;

$$\text{average delay per cycle} = \frac{1}{2}\frac{E\{R^2\}\lambda}{(1 - \lambda/\mu)}\left[1 + \frac{(I_A + I_D)E\{R\}}{(\mu - \lambda)E\{R^2\}}\right].$$

The correction term is still generally of minor importance.

The question of whether or not the queue clears in some time $R + G$ is of more practical interest. For given R and G, the distribution of the queue length at time $R + G, f_Q(l; R + G|R, G)$ is given by (9.36). Since the second term of (9.36) is of little consequence, however, we can approximate
$$P\{Q(R + G) > 0|R, G\} \simeq$$

$$1 - \Phi\left(\frac{(\mu - \lambda)(R + G - t_1)}{[(I_A\lambda + I_D\mu)(R + G - t_0)]^{1/2}}\right). \qquad (9.38)$$

The issue here is very similar to that described above for overloaded vehicles. For a fixed-cycle traffic signal, R and G are deterministic and the condition for the queue to clear nearly every cycle is that $R + G$ be larger than the time t_1 to clear the deterministic queue by an amount sufficient to serve one or two standard deviations of the queue length at rate $\mu - \lambda$, i.e., the argument of the Φ-function is about 1 or 2. For a typical traffic signal that can serve about 10 or 20 cars per cycle, overflow of the green will be fairly common if the arrival rate exceeds about 70 % of the maximum rate which can be served (for the given values of R and G).

If r_i and g_i vary from cycle to cycle so as to create a distribution of R and G one could have a situation analogous to that described above for buses in which the unconditional distribution of $Q(R+G)$ is determined primarily by the distribution of R and/or G, i.e., $P\{Q(R +G) > 0\} \simeq P\{R+G > t_1\}$ (almost) independent of the fluctuations of the arrivals or departures. This could conceivably happen at an intersection controlled by a vehicle-actuated signal for which the signal is controlled by traffic streams other than the one under consideration. Even then one would not expect $C^2(R)$ or $C^2(G)$ to be comparable with 1 as commonly occurs for the headways between buses.

The more common situation for traffic signals is that if r_i and g_i vary from cycle to cycle, they do so in response to the traffic fluctuations rather than external cuases. A vehicle-actuated signal typically will extend the green time until the queue vanishes (provided the green time does not exceed some predetermined maximum). It behaves analogously to a bus or elevator of arbitrarily large size which closes the doors when the queue has been served (loaded) and proceeds to do another task (take the passengers some place or for the traffic signal to serve another traffic stream).

9.6 Pulsed service with overflow

Suppose that any customers who are not served during one service pulse remain in queue until the next service pulse or as long as necessary to be served.

The queue length at any time does not depend upon the queue discipline but it is convenient to imagine that the queue is served last in first out. Any new arrivals during a cycle would then be served as described in the last section independently of any residual queue left from the previous cycle. The total delay to all customers during a cycle would be the sum of the delay to the new arrivals as evaluated above plus the delay to the residual queue.

For buses or elevators with a negligible loading time any customer left from a previous cycle will wait a time R (in one cycle). If a cycle starts a time 0 with a residual queue $Q(0)$ the total wait for these customers during this cycle is $RQ(0)$. For a process that continues for many cycles the arithmetic average of any quantity over many cycles will be denoted by $E\{\cdot\}$. The average delay per cycle for these customers is thus represented as $E\{RQ(0)\}$. The corresponding average queue length over many cycles is $E\{RQ(0)\}/E\{R\}$. If the time R is statistically independent of the residual queue from the previous cycle $E\{RQ(0)\} = E\{R\}E\{Q(0)\}$ and the average queue length over many cycles due to the residual customers is $E\{Q(0)\}$.

For a traffic signal, any customer in the residual queue at time 0 that is not served in the next cycle will wait a time $R + G$ during that cycle. Any customer in the residual queue that is served during that cycle will be served during the time after the new arrivals have been served until the end of the green time. They will be delayed for a time less than $R + G$. For any arrival rates that would cause a residual queue to form, however, most of the green time will be spent serving the new arrivals. Any saving in delay to the residual queue because it is served before time $R + G$ is expected to be of little importance. It suffices to assume that all customers in the residual queue are delayed the full cycle $R + G$. The average residual queue over many cycles is therefore approximately $E\{(R + G)Q(0)\}/E\{R + G\}$. If $R + G$ is statistically independent of $Q(0)$, or if $R + G$ is deterministic, the average residual queue is simply $E\{Q(0)\}$.

To analyze the consequences of an overflow of the queue from one cycle to the next it suffices, for all practical purposes, to study the properties of the queue only at the end of service pulses, i.e., $Q(0)$, $Q(R + G)$, etc.

The evolution of $Q(0)$, $Q(R + G)$, . . . is essentially that of a 'bulk service queue'. Suppose for the loading of buses we let

$$Y = (\text{number of arrivals in time } R) - M$$

and for the traffic signal

$$Y = (\text{number of arrivals in time } R + G) - (\text{number of customers who could be served in time } G \text{ if the server were busy at all times})$$

i.e., Y represents the potential change in $Q(t)$ in one cycle (if $Q(t)$ could become negative) and

$$Q(R + G) = \max\{0, Q(0) + Y\}.$$

For buses the distribution of Y corresponding to (9.31) is

$$1 - F_Y(y) = \int_0^\infty \left[1 - \Phi\left(\frac{y + M - \lambda t}{(I_A \lambda t)^{1/2}} \right) \right] f_R(t) dt \qquad (9.39)$$

i.e., Y given R is (approximately) normal with mean $\lambda R - M$ and variance $I_A \lambda R$.

Correspondingly for the traffic signal Y given R and G is approximately normal with mean $\lambda(R + G) - \mu G$ and variance $I_A \lambda \times (R + G) + I_D \mu G$. The unconditional distribution is the expectation of this normal distribution over the joint distribution of R and G.

If the new arrivals in one cycle are statistically independent of the residual queue, $Q(0)$, then $Q(0)$, $Q(R + G)$, ... define a Markov process with

$$P\{Q(R + G) \le x\} = \begin{cases} \displaystyle\int_0^\infty P\{Q(0) < z\} f_Y(x - z) dz & \text{for } x \ge 0 \\ 0 & \text{for } x < 0, \end{cases}$$

If there is an equilibrium distribution such that

$$P\{Q(R + G) \le x\} = P\{Q(0) \le x\} \equiv F_Q(x)$$

then $F_Q(x)$ satisfies the integral equation

$$F_Q(x) = \begin{cases} \displaystyle\int_0^\infty F_Q(z) f_Y(x - z) dz & \text{for } x \ge 0 \\ 0 & \text{for } x \le 0. \end{cases} \qquad (9.40)$$

Equations of this type are known as Weiner–Hopf integral equations. Solutions in the form of integral representations can be obtained by Fourier integral and complex variable methods but these formal solutions are generally quite unwieldy. It is possible, however, to obtain fairly simple approximate solutions if a positive (residual) queue occurs only rarely or if the queue typically persists for many cycles.

If there is seldom a queue it is advantageous to write (9.40) as

$$F_Q(x) = F_Y(x) - \int_0^\infty [1 - F_Q(z)] f_Y(x - z) dz$$

or

$$[1 - F_Q(x)] = [1 - F_Y(x)] + \int_0^\infty [1 - F_Q(z)] f_Y(x - z) dz. \qquad (9.41)$$

As a zeroth approximation to the solution of (9.41) we take

$$1 - F_Q^{(0)}(x) = \begin{cases} 0 & x > 0 \\ 1 & x \le 0 \end{cases}$$

i.e., $Q = 0$ with probability 1. For a first approximation we substitute $F_Q^{(0)}(x)$ for $F_Q(x)$ into the right hand side of (9.41) and write

$$[1 - F_Q^{(1)}(x)] = [1 - F_Y(x)] \quad \text{for} \quad x > 0. \tag{9.42}$$

This is actually the queue distribution that would exist at time $R + G$ given that $Q(0) = 0$. This approximation, in effect, implies that a queue distribution which is close to equilibrium at time 0 will be even closer to an equilibrium at time $R + G$. A queue occurs rarely, of course, if $1 \gg [1 - F_Y(0)] > [1 - F_Y(x)]$, i.e., nearly all customers who arrive in a cycle are served in that cycle.

The next or any succeeding approximations are obtained by substituting the previous approximation into the right hand side of (9.41). For example,

$$1 - F^{(2)}(x) = 1 - F_Y(x) + \int_0^\infty [1 - F_Y(z)] f_Y(x - z) dz. \tag{9.43}$$

Succeeding approximations will involve multiple convolutions of the distribution $f_Y(\cdot)$, which become progressively more tedious to evaluate numerically.

Another equivalent way of describing the above sequence of approximations is to note that if Q_j is the residual queue on the jth cycle and Y_j the Y for the jth cycle, then

$$Q_j = \max \{0, Q_{j-1} + Y_{j-1}\}.$$

If we iterate this relation

$$\begin{aligned} Q_j &= \max \{0, \max \{0, Q_{j-2} + Y_{j-2}\} + Y_{j-1}\} \\ &= \max \{0, Y_{j-1}, Q_{j-2} + Y_{j-1} + Y_{j-2}\} \\ &= \max \{0, Y_{j-1}, Y_{j-1} + Y_{j-2}, Y_{j-1} + Y_{j-2} + Y_{j-3}, \ldots\}. \end{aligned} \tag{9.44}$$

The successive approximation described above correspond to taking the maximum of only 1, 2, 3, etc., of these terms.

If the Y_j are independent and $E\{Y\} < 0$,

$$\sum_{k=1}^n Y_{j-k}$$

will become approximately normally distributed for $n \gg 1$ with mean

$nE\{Y\}$ and standard deviation $[n\,\text{Var}\,\{Y\}]^{1/2}$. For sufficiently large n, specifically for $|nE\{Y\}|/[n\,\text{Var}\,\{Y\}]^{1/2} \gg 1$, there will be a negligible probability that

$$\sum_{k=1}^{n} Y_{j-k} > 0$$

in (9.44). The above scheme of successive approximations will converge to the equilibrium distribution provided $E\{Y\} < 0$ (the obvious condition for the system to be undersaturated). It will typically converge quite rapidly if $E\{Y\}/[\text{Var}\,\{Y\}]^{1/2} \gtrsim 1$.

If, on the other hand, $E\{Y\}/[\text{Var}\,\{Y\}]^{1/2} \ll 1$, we anticipate that $E\{Q\}$ will become large compared with $[\text{Var}\,\{Y\}]^{1/2}$ and that the typical fractional change in Q from one cycle to the next will be small compared with 1. The sequence of Q_j will behave like a diffusion process with a reflecting barrier. Specifically, the distribution function $F(x; j)$ of Q_j will satisfy approximately the diffusion equation

$$\frac{\partial F(x; j)}{\partial j} = \left[-E\{Y\}\frac{\partial}{\partial x} + \frac{\text{Var}\,\{Y\}}{2}\frac{\partial^2}{\partial x^2} \right] F(x; j),$$

with mean drift per cycle $E\{Y\}$ and variance rate (per cycle) of $\text{Var}\,\{Y\}$. The equilibrium distribution will, therefore, be approximately

$$1 - F_Q(x) \simeq \exp\left[\frac{2E\{Y\}}{\text{Var}\,\{Y\}} x \right],$$

with

$$E\{Q\} = -\frac{\text{Var}\,\{Y\}}{2E\{Y\}}.$$

Bibliography

Books on queueing theory in English

1948

Brockmeyer, E., Halstrom, H. A. and Jensen, A., *The Life and Works of A. K. Erlang*, Danish Academy of Technical Science.

1958

Morse, P., *Queues, Inventories, and Maintenance*, John Wiley, New York.
Peck, L. G. and Hazelwood, R. N., *Finite Queueing Tables*, John Wiley, New York.

1960

Khintchine, A. Y., *Mathematical Methods in the Theory of Queueing*, Griffin, London.
Syski, R., *Introduction to Congestion Theory in Telephone Systems*, Oliver and Boyd, London.

1961

Cox, D. R. and Smith, W. L., *Queues*, Chapman and Hall, London.
Saaty, T. L., *Elements of Queueing Theory with Applications*, McGraw–Hill, New York.

1962

Descloux, A., *Delay Tables for Finite- and Infinite-Source Systems*, McGraw–Hill, New York.
Riordan, J., *Stochastic Service Systems*, John Wiley, New York.
Takács, L., *Introduction to the Theory of Queues*, Oxford University Press, New York.

1963

Beneš, V. E., *General Stochastic Processes in the Theory of Queues*, Addison–Wesley, Reading, Mass.

1965

Beneš, V. E., *Mathematical Theory of Connecting Networks and Telephone Traffic*, Academic Press, New York.
Prabhu, N. U., *Queues and Inventories*, John Wiley, New York.
Smith, W. L. and Wilkinson, W. E. (eds), *Proceedings of the Symposium on Congestion Theory*, University of North Carolina Press, Chapel Hill.

1966

Lee, A. M., *Applied Queueing Theory*, St. Martin's Press, Montreal.

1967

Cruon, R. (ed.), *Queueing Theory, Recent Developments and Applications*, Elsevier, New York.
Ruiz-Pala, E. Avila-Beloso, C. and Hines, W. W., *Waiting-Line Models*, Reinhold, New York.

1968

Beckman, P., *Introduction to Elementary Queuing Theory*, Golem Press, Boulder, Colorado.
Bhat, U. N., *A Study of the Queueing Systems M/G/1 and GI/M/1*, Lecture Notes in Operations Research and Mathematical Economics # 2, Springer–Verlag, Berlin.
Gnedenko, B. and Kovalenko, J., *Introduction to Queueing Theory*, Israel Program for Scientific Translations, Ltd, Jerusalem.
Jaiswal, N. K., *Priority Queues*, Academic Press, New York.

1969

Cohen, J. H., *The Single Server Queue*, North Holland Publishing Co., Amsterdam.
Khintchine, A. Y., *Mathematical Methods in the Theory of Queueing*, 2nd edn, Hafner Publishing Co., New York.
Panico, J., *Queuing Theory: A Study of Waiting Lines for Business, Economics and Science*, Prentice–Hall, Englewood Cliffs, N. J.

1970

Ghosal, A., *Some Aspects of Queueing and Storage Systems*, Springer–Verlag, Berlin.

1971

Newell, G. F., *Applications of Queueing Theory*, Chapman and Hall, London.

1972

Bagchi, T. P., and Templeton, J. G. C., *Numerical Methods in Markov Chains and Bulk Queues*, Lecture Notes in Operations Research and Mathematical Economics #72, Springer–Verlag, Berlin.
Cooper, R. B., *Introduction to Queueing Theory*, Macmillan, New York.
Page, E. G., *Queueing Theory in OR*, Crane Russak, New York.

1973

Newell, G. F., *Approximate Stochastic Behavior of n-Server Service Systems with Large n*, Lectures Notes in Economic and Mathematical Systems #87, Springer–Verlag, Berlin.
Kosten, L., *Stochastic Theory of Service Systems*, Pergamon, London.

1974

Clarke, A. B., (ed), *Mathematical Methods in Queueing Theory*, Lecture Notes in Economic and Mathematical Systems #98, Springer–Verlag, Berlin.
Gross, D. and Harris, C. M. *Fundamentals of Queueing Theory*, John Wiley.

1975

Conolly, B., *Lecture Notes on Queueing Theory*, Chichester, Ellis Horwood.
Kleinrock, L., *Queueing Systems*, Vol. 1, Wiley–Interscience, New York.

1976

Kleinrock, L., *Queueing Systems*, Vol. 2, *Computer Applications*, Wiley–Interscience, New York.

1979

Allen, A. O., *Probability, Statistics, and Queueing Theory with Computer Science Applications*, Academic Press.

Newell, G. F., *Approximate Behavior of Tandem Queues*, Lecture Notes in Economics and Mathematical Systems #171, Springer–Verlag, Berlin.

1980

Bruell, S. C., and Balbo, G., *Computational Algorithms for Closed Queueing Networks*, North Holland.
Cohen, J. H., *The Single Server Queue*, 2nd edn, North Holland Publishing Co., Amsterdam.
Kelly, F. P., *Reversibility and Stochastic Networks*, Wiley, New York.

1981

Cooper, R. B., *Introduction to Queueing Theory*, 2nd edn, Elsevier, North Holland, New York.
Franken, P., Konig, D., Arndt, U. and Schmidt, K., *Queues and Point Processes*, Akademic Press.
Hillier, F. S. and Yu, O. S., *Queueing Tables and Graphs*, Elsevier, North Holland, New York.

Deterministic queueing models

Barbo, W. A., (1967) The use of queueing models in design of baggage claim areas in airports. ITTE Graduate Report, University of California.
Bavarez, E. and Newell, G. F. (1967) Traffic signal synchronization on a one-way street. *Transportation Science*, 1, 74–80.
Browne, J. J., Kelly J. J. and Le Bourgeois, P. (1970) Maximum inventories in baggage claim: a double ended queuing system. *Transporation Science*, 4, 64–78. Erratum, *ibid.*, 4, 327.
Chu, K. C. (1977) Decentralized real-time control of congested traffic networks. *Proc. Seventh Int. Symp. on Transportation and Traffic Theory*, Kyoto.
Chu, K. C. and Gazis, D. C. (1974) Dynamic allocation of parallel congested traffic channels. *Transportation and Traffic Theory (Proc. Sixth Int. Symp. on Transportation and Traffic Theory)*, A. H. and A. W. Reed, Sydney.
D'Ans, G. C. and Gazis, D. C. (1976) Optimal control of oversaturated store-and-forward transportation networks. *Transportation Science*, 10, 1–19.
Dunne, M. C. and Potts, R. B. (1964) Algorithm for traffic control. *Operations Research*, 12, 870–881.
Edie, L. C. and Baverez, E. (1967) Generation and propagation of stop-start traffic waves. *Vehicular Traffic Science (Proc. Third Int. Symp. on the Theory of Traffic Flow)*, American Elsevier, N. Y., 26–37.
Edie, L. C. and Foote, R. S. (1961) Experiments on single-lane flow in tunnels. *Theory of Traffic Flow*, Elsevier, Amsterdam, 175–92.

Edie, L. C. and Foote, R. S. (1960) Effect of shock waves on tunnel traffic flow. *Proc. Highway Research Board*, **39**, 492–504.

Foulkes, J. D., Prager, W. and Warner, W. H. (1954) On bus schedules. *Management Science*, **1**, 41–8.

Grafton, R. B. and Newell, G. F. (1967) Optimal policies for the control of an undersaturated intersection. *Vehicular Traffic Science (Proc. Third Int. Symp. on the Theory of Traffic Flow)*, American Elsevier, N. Y., 239–57.

Gazis, D. C. (1964) Optimal control of a system of oversaturated intersections. *Operations Research*, **12**, 815–31.

Gazis, D. C. (1967) Optimal assignment of a reversible land in an oversaturated two-way traffic link. *Vehicular Traffic Science (Proc. Third Int. Symp. on the Theory of Traffic Flow)*, American Elsevier, N. Y., 181–90.

Gazis, D. C. and Potts, R. B. (1965) The oversaturated intersection. *Proc. Second Int. Symp. on the Theory of Traffic Flow*, OECD, Paris, 221–37.

Haji, R. and Newell, G. F. (1971) Optimal strategies for priority queues with nonlinear costs of delays. *SIAM J. Applied Mathematics*, **20**, 224–40.

Hurdle, V. F. (1973) Minimum cost schedules for a public transportation route. I: Theory. II: Examples. *Transportation Science*, **7**, 109–57.

Hurdle, V. F. (1974) The effect of queueing on traffic assignment in a simple road network. *Transportation and Traffic Theory (Proc. Sixth Int. Symp. on Transportation and Traffic Theory)* A. H. and A. W. Reed, Sydney, 519–40.

Lighthill, M. H. and Whitham, G. B. (1955) On kinematic waves. II: A theory of traffic flow on long crowded roads. *Proc. Royal Soc. (London)*, **A229**, 317–5.

Makigami, Y., Newell, G. F. and Rothery, R. (1971) Three-dimensional representations of traffic flow. *Transportation Science*, **5**, 302–13.

May, A. D. and Keller, H. E. M. (1967) A deterministic queueing model. *Transporation Research*, **1**, 117–28.

Moskowitz, K. and Newman L. (1963) Notes on freeway capacity. *Highway Research Record #27*, Highway Research Board.

Newell, G. F. (1969) Properties of vehicle-actuated signals. I: One-way streets. *Transportation Science*, **3**, 30–52.

Newell, G. F. (1969) Traffic signal synchronization for high flows on a two-way street. *Beitrage zur Theorie des Verkehrsflusses (Proc. Fourth Int. Symp. on the Theory of Traffic Flow)*, 87–92.

Newell, G. F. (1971) Dispatching policies for a transportation route. *Transportation Science*, **5**, 91–105.

Newell, G. F. (1973) *Approximate Stochastic Behavior of n-Server Service Systems with Large n*. Lecture Notes in Economics and Mathematical Systems #87, Springer–Verlag.

Newell, G. F. (1977) The effect of queues on the traffic assignment to freeways. *Proc. Seventh Int. Symposium on Transportation and Traffic Theory*, Kyoto, 311–40.

Newell, G. F. (1979) *Approximate Behavior of Tandem Queues.* Lecture Notes in Economic and Mathematical Systems #171, Springer–Verlag.

Newell, G. F. (1979) Airport capacity and delay. *Transportation Science*, **13**.

Newman, L. (1963) Reduction in delay by increasing bottleneck capacity. *Traffic Engineering*, Nov., 28–32.

Oliver, R. M. and Samuel, A. H. (1962) Reducing letter delays in post offices. *Operations Research*, **10**, 839–92.

Rangarajan, R. and Oliver, R. M. (1967) Allocation of servicing periods that minimize average delay for *N* time-shared traffic streams. *Transportation Science*, **1**, 74–80.

Salzborn, F. J. M. (1972) Optimum bus scheduling. *Transportation Science*, **6**, 137–48.

Yaffe, H. J. (1974) A model for optimal operation and design of solid waste transfer stations. *Transportation Science*, **8**, 265–306.

Author index

Subject index

Absorbing boundary, 232
Airports
    baggage, 97, 98, 103, 296
    gate occupancy, 18, 180, 212, 241
    passenger arrivals, 112
    runways, 79, 80, 103, 122, 123, 140, 213, 277, 298
Arrival rate, 2, 3, 13, 17–19, 111
Arrival times, 4, 6, 11, 13, 21, 105, 108, 109, 111
Average
    arrival rate, 13, 15, 18, 60
    cost of delay, 16
    cumulative arrivals, 29, 30
    headway, 46
    number in service, 18–19
    queue length, 14, 17, 18, 29, 44, 108, 109
    service rate, 13, 32
    service time, 17–19
    time in system, 17
    wait in queue, 14, 17, 44, 108, 109

Batch arrivals, 123–4, 141, 192–3
Binomial distribution, 112, 180, 184–5
Blocking, 66, 68, 164–5, 240
Boundary conditions, 215, 217, 225, 231–4
Brownian motion, 127, 137, 217, 222, 227, 232
Bulk server, 12, 72, 282, 288–91, 295
Bus dispatching, 48, 51–2, 72, 75, 281–3, 287, 297–8

Capacity, 66–69, 282
Central limit theorem, 119, 125, 128, 131
Coefficient of variation, 46

Compound Poisson process, 123–4, 129
Conservation equation, 60, 223, 233
Conservation principle, 1–2, 4–5, 53, 72, 80
Continuum of servers, 58–60
Cost of delay, 3–4, 23, 48, 50, 79–80, 84–5, 91–7
Covariance, 148
Cumulative
    arrivals, 6–7, 40, 55, 80, 107
    departures, 7, 9, 11, 55, 57, 80, 107
    number of servers, 73
    value, 6, 80–1
    work, 7, 81–4, 92, 136
Customers, 4

Demand for service, 39
Delay, 3, 4, 105
Departure
    rate, 32
    times, 8–10, 13, 21
Deterministic fluid approximation, 25–49, 126, 225, 273, 296–8
Diffusion
    equations, 215–91
    process, 127–9
Distribution function, 110

Elevators, 72, 75, 102, 194, 281–2, 287
Embedded Markov chain, 168
Emperical distribution function, 110–3
Equilibrium queue distributions, 143–74, 195–211, 237–44
Ergodic property, 146, 148, 150
Erlang loss function, 196
Expectation, 107